Robin Dunbar is Professor of Ev... ...versity of Liverpool and has held .... Universities of Cambridge and Stockholm. He has been praised for 'writing that is dizzyingly multidisciplinary but shows great generosity to the ordinary reader' (*Guardian*). His books include *The Trouble with Science* (1995), 'an eloquent riposte to the anti-science lobby' (*Sunday Times*), and *Grooming, Gossip and the Evolution of Language* (1996), praised as 'brilliantly original' and 'a delight to read' (*Focus*).

His main research interests are the evolution of the mind, and the social systems of human and non-human primates; he has carried out field studies of monkeys and antelope in East and West Africa, and of wild goats in Scotland. In June 2003 he led a team of academics which won the largest single grant ever awarded by the British Academy, to research what it means to be human.

Praise for ***Grooming, Gossip and the Evolution of Language***:

'A modern classic of pop science.' *Guardian*

'I learnt something fascinating on pretty much every page of this impressively multidisciplinary book.' *Independent on Sunday*

Praise for ***The Trouble with Science***:

'A terrific book.' *New Scientist*

'A strong but accessible defence of science . . . It deserves to be widely read – not least by journalists and the new academic critics of science.' *Nature*

'Just the sort of science that every educated person ought to know about.' *Independent on Sunday*

by the same author

GROOMING, GOSSIP AND THE EVOLUTION OF LANGUAGE
THE TROUBLE WITH SCIENCE

ROBIN DUNBAR

The Human Story

A new history of mankind's evolution

faber and faber

First published in 2004
by Faber and Faber Limited
74 77 Great Russel Street, London WC1B 3DA
This paperback edition first published in 2005

Typeset by Faber and Faber Ltd
Printed and bound by CPI Group (UK) Ltd, Croydon, CR0 4YY

A CIP record for this book
is available from the British Library

ISBN 978–0–571–22303–9

10 9 8 7 6 5 4

Notes on drawings for chapter headings
(*all drawings by Arran Dunbar*)

1 Bison: Altamira cave, northern Spain, approx. 12,000 years BP
2 San rock art: Drakensberg Mountains, South Africa (after D. Lewis-Williams, 2002, *A Cosmos in Stone*, Altamira Press, Walnut Creek)
3 Stencilled hands: Fuente del Salin cave, northern Spain, approx. 20,000 years BP
4 Three lionesses: Chauvet cave, south-east France, c. 30,000 years BP
5 Eland-headed dancers: San rock art, Medikane cave, Lesotho, southern Africa (after D. Lewis-Williams, 2002, *A Cosmos in Stone*, Altamira Press, Walnut Creek)
6 Venus of Lespugue: Lespugue cave, south-west France, approx. 25,000 years BP
7 Dancing therianthrope (shamanic animal-headed figure): Volpe cave, southern France, approx. 12,000 years BP (after D. Lewis-Williams, 2002, *The Mind in the Cave*, Thames and Hudson, London)

For Steve
who set me on my way

Emancipate yourselves from mental slavery,
None but ourselves can free our mind . . .
Bob Marley, *Redemption Song* (1980)

Contents

1 Visions in Stone

The light from the tallow lamp flickered uncertainly in a draught, breaking the man's concentration. He stood back for a moment to survey his handiwork. He picked up the carved stone bowl that served as a lamp and held it up to the rock wall, the better to see where he had been working. Around him on the wall, images of animals seemed to move, magically brought to life by the guttering tallow wick. A great jumble of bison, deer, and horses cascaded through a timeless space. Here, one seemed caught in a moment of surprise; there a bison rested on its haunches with its head turned as though to survey the unwarranted intrusion into its thoughts as it lay quietly chewing the cud.

He set to work again, his brow puckered in concentration, carefully drawing in the outlines of a new animal with a piece of charcoal. Warmed by the fire of his imagination, he was oblivious of the coldness of the cave around him. In his mind's eye, he saw what he had to draw, a vision of a world as real and tangible as the stone on which he worked – a place where animals gambolled through woodlands or grazed in grassy glades, sunlight dappling their shoulders. He worked with all the skill he could muster, intent on capturing the vision in his mind before it vanished.

He had spent years of his life perfecting these skills, coming back to this cave every spring, when the land was bursting with new life, to create his own imaginary life here on the rock walls. Sometimes, he came alone; sometimes, with others. But always with the same goal: to undertake the journeys, and then to make an enduring record of his travels, to capture the sights he saw, the wrenching emotions he felt as he sped through dark and often dangerous places – journeys whose end he could never quite foretell.

He always knew when that moment had arrived, though he could not always anticipate it, for it was as elusive as the deer that he hunted in the woods. Even though he had set out on the journey many times, the great mystery was that the path always seemed to be different. Only the end, when it came, was the same, bringing that sense of arrival at a familiar place, that feeling of relief tinged with exhaustion from a difficult road travelled, of dangers that had been safely negotiated once more.

And now, as after every journey, he came back to the cave to record his experiences and what he had seen. The bison's form on which he was working gradually took shape the way he had seen it – so close up that, when it had turned its great head towards him and fixed him languidly with one eye, it was as though he was peering into its mind through that great liquid portal. Now, its eye gazed back at him from the wall, fixing him once again with its glassy stare. In his mind, he could recreate that moment of surprise, of fear, that always accompanied the uncertainty when he could not be sure what the great beast would do. Sometimes, by turning away to the darkness of the cave and then turning suddenly back again – or by coming unexpectedly upon some earlier painting that he had forgotten about – he could sense once more that same tension just for a fleeting moment.

So engrossed was he in his work that several hours passed without his noticing. But eventually, the tiredness in his arms and the growing pangs of hunger persuaded him to put down his drawing materials and set off down the long winding passage back to the mouth of the cave. He emerged into the harsh light of late after-

noon from the cave's mouth. The entrance was quite deceptive, being hidden behind a jumble of boulders and beech trees. It gave no clue to the size of the cave behind – the long winding tunnels that you sometimes had to crawl along on your belly, the vaulted chambers that seemed to occur with no plan along the tunnel itself or opened off as blind side-chambers.

He glanced at the sun hanging above the western horizon and saw that it was already late afternoon. He blew out the wick in the tallow lamp and placed it in a small niche in the rock wall just inside the entrance. And, feeling content with his day's work, set off down through the trees to where their camp nestled beside the river in the valley bottom a few miles downstream.

One hundred and eighty centuries later, in 1879, Maria, the young daughter of Don Marcelino Sanz de Sautuola, idly glanced up towards the cave ceiling above her head while her father busied himself beside her, searching the floor for prehistoric artefacts. What she saw in the faint glow of the oil lamp that her father had placed beside him made her reach involuntarily to grab onto his coat-tail: looming down at her out of the gloom, bison and horses were pouring out of the rock. Disturbed from his searches, Don Marcelino turned irritably to remonstrate with the child. But her up-turned face – her eyes fixed on something above him, her mouth wordlessly opening and closing – at once made him realise that there was something unusual. He looked slowly up above him, his eyes searching the gloom. He reached for the oil lamp and held it above his head so that he could see better . . . and let out a gasp. Above him, the bison, deer and horses turned and twisted, bunched up against each other, fighting for space, or lay chewing the cud, just as they had been left 18,000 years ago by the painters who had made them.

For Don Marcelino, this was a discovery beyond compare.

Excited and impressed by the carved statuettes and ivory plaques that had been found in the recently discovered prehistoric caves in nearby southern France, he had spent several years searching the caves near Santander in northern Spain in the hope of discovering his own treasure trove of prehistoric art. His efforts had proved fruitless. And now, his searches in the cave at Altamira had accidentally led him to the discovery of some extraordinarily spectacular prehistoric paintings. His reputation was surely made. The rich and the famous, the scholars and the antiquarians, would flock to his cave and he would be fêted at all the great scientific gatherings for years to come.

Don Marcelino was to die a disappointed man. After an initial flurry of interest, the antiquarians of the day declared the paintings too advanced to be the work of primitive man. Rather, they must have been painted by someone who had visited the cave in recent years . . . perhaps Don Marcelino himself. Though they never quite accused him of forgery to his face, the sniff of it seemed to be there in the fetid air of the Altamira cave. Don Marcelino withdrew to his family estate. He died, a frustrated and embittered man, just nine years later. It was not until 1902 that the paintings were accepted as being of genuine prehistoric age. More extensive explorations had eventually revealed that the cave had gallery after gallery of paintings, sketches and drawings running more than 200 metres into the hillside. But by then, Don Marcelino had been in his grave the better part of twenty years, and his grown-up daughter had other things to distract her from the paintings that had so startled her as a child that summer morning.

Yet the question of who might have painted these extraordinary visions, or why, remained an enduring mystery.

The cave at Altamira, it now seems, is far from unique: there are around 150 known sites of prehistoric cave art in Europe.

Although examples are known from as far east as the Ural Mountains in Russia and one has recently even been discovered in England, all but a handful are concentrated in southern France and the Spanish peninsula. Something about the caves in this area or about the particular human groups that occupied this area between 25,000 and 12,000 years ago combined to make cave art especially rewarding. The artwork itself is little short of exquisite. It is easy, in the dark of these caves, to become lost in the mystery of the figures that some unseen hand sketched so beautifully so long ago. Grown men have been reduced to tears before them.

Here, in one corner of an ancient gallery, is a child's hand, stencilled around by paint blown from the mouth. If the guardians of the cave would allow it,* you could place your own hand over the outline, and reach out across the millennia, as it were, to touch that child. A delicate, hesitant touch, such as one might give to a new lover. It is impossible not to feel the magic in the air. Who was he – or was it she? By what name was he or she known? And what became of this child? Did he or she grow up, have children of his or her own, and live to a ripe old age, a respected white-haired member of the community, remembering a day – in spring, perhaps – when he or she had been led down the winding tunnels by the dim light of a tallow lamp to a remote back chamber and made to press a hand against the cold wall of the cave while one of the men blew paint across it? Or, instead perhaps, did the child die in infancy, of illness or accident, or fall prey to a wandering predator – a future cut off in the first flush of childhood, one of many small tragedies in

---

* Needless to say, touching the images on the cave walls is strictly prohibited, for after even a few such contacts the fragile paint would be worn away and the paintings lost. Indeed, even the breath of the thousands of visitors that crowded through the caves during the decades after their first discovery was enough to introduce bacteria that began to eat away at the paint. Many caves are now barred to the public, who may instead view replicas nearby.

the life of its mother, its passing signalled by a shrill, brittle wail of inconsolable grief?

What we know is that the people that made these drawings engaged life with an exuberance that resonates with us today. Cave art is the final flowering of a remarkable development in human evolutionary history, a phenomenon that archaeologists refer to as the Upper Palaeolithic Revolution. It began around 50,000 years ago with a sudden burst of very much more sophisticated stone, bone and wooden tools – including needles, awls, fishhooks, arrow- and spearheads. From around 30,000 years ago, this is followed up by a veritable explosion of artwork that has no particular function in terms of everyday survival but seems rather to be entirely decorative. There are brooches, carved buttons, dolls, toy animals and, perhaps most spectacular of all, figurines – exemplified above all by the so-called Venus figures of central and southern Europe. These famous 'Michelin-tyre' ladies seem to have been the pin-ups of their day. Big-hipped and ample-bosomed, with their hair often beautifully braided, these ivory and stone (sometimes even baked clay) statuettes are quite the most striking of the late Palaeolithic artefacts. Then, from about 20,000 years ago, we begin to find evidence for deliberate burials, for music and for a life in the mind. The cave paintings of Altamira, Lascaux, Chavette and the many other grottoes, shelters and caverns across southern Europe and beyond are but the icing on this grand artistic cake. Nothing like it had ever been seen in the history of human evolution. Buried within it lay the foundations for modern human culture, from literature to religion and, beyond, to science.

This outpouring of craftsmanship speaks to us across the intervening millennia. Here is a people not so very different from ourselves: what we find beautiful, they too found beautiful. Here, it seems, encapsulated in a brief moment in time, is the essence of what made us who we are, what finally produced humans as we know them, with all that inflorescence of culture that makes us in some intangible but very certain way utterly different from every other species alive today – and, indeed, every other species that preceded us in the long history of life on earth.

But who are we, this species of painters and poets? How did we come to be here? How was it that these nameless cave painters of southern Europe came to ply their trade there so long ago? Where did they come from? And why was it only they, of all the species that have ever been, who had the wit to leave their delicate imprint behind them? And, why – perhaps most intriguing of all – why did they do it?

This book is an odyssey, a journey up through the mists of time from the remote past. It explores what must perhaps be the most fundamental of all questions – who we are. What is it that sets us so firmly apart from all those other species with whom we share the planet? How – given that, at conception, ours is the very same beginning as that of every other life form – do the differences between us and all other species come about during the course of human life? When in the course of our evolutionary history did those differences that separate us from our fellow creatures come to be? And, maybe the most tantalising question of all, *why* did these differences come to bless our lineage and no other?

This is a journey within ourselves. To understand what it is to be human, we have to understand our own minds. It is here, in our ability to reflect upon ourselves and our relationship with the world 'out there', that the real differences between us and the rest of creation seem to lie. Our physical attributes and a great deal of our behaviour are unexceptional, even by the standards of an unexceptional group like the primates. Rather, what sets us apart is, above all, a life in the mind, the ability to imagine. As obvious as this may seem, it is only very recently that we have been able to pinpoint exactly what these features of the mind are that set us apart. So much of what we do is similar to what we see in our monkey and ape cousins, their inventiveness and intelligence, their intensely social ways of life, even their remarkable evolutionary success as a group. Yet we remain apart from them, distanced by that indefinable mental world that we claim as our own.

In exploring this world, we shall have to draw on many different sciences, each of which will give us only partial answers. The past decade or so has seen astonishing advances in many disciplines, from genetics to behavioural studies to psychology. We are still absorbing their findings and coming to terms with their implications. In their different ways, they have so revolutionised our understanding of who we are that our view of ourselves – and, in turn, how we view the other species with whom we share our past as well as our future – has been turned topsy-turvy. Only by drawing together these many disparate threads will we be able to come to some real understanding of just what it is that makes us who we are.

Our history has been a long one. In one sense, it began some sixty-five million years ago when the dinosaurs trod the steaming tropical forests of Europe and North America in undisputed mastery of the planet. Our earliest ancestors, as yet barely recognisable as primates, skittered through the trees and bushes much as squirrels do today. Later, in the aeons after the dinosaurs passed into that great dinosaur Valhalla, these primitive squirrel-like mammals diversified and evolved into a highly successful group of animals. They became the ancestors of the monkeys and apes with which we are now so familiar.

Much later, some six to seven million years ago, one of their many descendants began to develop some new characteristics and a slow but steady divergence developed between their lineage and that of the other African apes – the chimpanzees and the gorillas. At first, these evolutionary innovations involved a handful of rather uninteresting features, mostly related to bipedal walking. But eventually, genuinely novel features began to appear in this lineage – a rapidly enlarging brain, tool use, language, culture. That lineage ultimately gave rise to our cave artist, and, a little later still, to us, modern humans. The road that led from our common ancestor with the African apes some six million years ago to us was an uncertain one, dogged by serendipity and catastrophes that sent us spinning down unexpected evolutionary pathways. There was no certain sequence of changes leading inexorably from apes to humans with some God-given inevitability; there was only the eternal chaos of evolutionary history.

So, let us imagine ourselves in the unfamiliar environment of a wooded plain in eastern Africa around three and a half million years ago. It is mid-afternoon, and the sun is beginning its gradual descent towards the horizon. In the distance, the shimmering heat haze condenses into a number of human-like shapes straggling their way across the wooded landscape.

2 The Ape on Two Legs

In the sultry warmth of an African savannah, tiny dust devils, stirred up by light winds that began nowhere and were too ephemeral to survive long enough to go anywhere, skipped nervously across the landscape. Beyond them, the volcanic cone of what would eventually come to be known as Mount Sadiman loomed over the plain, grumbling fitfully. Had they been more alert, the dozen or so figures working their way steadily across the plain towards a grove of trees near the mountain's foot might have thought the better of spending a day in the open. But the day had begun no differently from any other, and they were used to the mountain's tetchy serenade of intermittent rumbles. Oblivious of what might lie in store, they walked steadily on.

Then, deep within the earth's crust, a particularly large bubble of hot acid detached itself from the mountain's bowels and burst through the outer crust to spew hot ash, fumes and lava into the atmosphere. The plume of smoke and ash that rose tens of thousands of feet into the air in a matter of seconds drifted down in a rain of black dust that settled everywhere for miles around. The little band paused, turning as one to stare at the mountain across the plain.

The rumblings and ash-falls continued throughout the afternoon. Each explosion deep within the volcano, each burst of flame-lit ash cloud that rose above the crown, each surge of glowing lava that poured in searching tongues down the volcano's side, set off paroxysms of panic among the wildlife in the plain below. The night that followed was especially disturbed for the little band, as they huddled in the branches of the trees down in a riverbed. The mountain's feverish groanings kept most of them awake. Those that dozed off in the lulls were soon woken by the whimperings of the youngsters.

When morning dawned, they peered uncertainly through the drifting acrid mists, wondering which way to go. From these particular trees, their normal route would have taken them directly towards the mountain's slopes where they knew of an especially good patch of fig trees whose fruit crop would now be just beginning to ripen. But the mountain's uncharacteristic activity did not inspire much enthusiasm among them for that course. As the sun began to warm the chill landscape, they clambered down one by one to the ground from their night-time refuges. Several of them began to pick at the small green fruits that covered low bushes nearby. Others squatted on the ground, staring at the mountain. A few picked at the grey powdery ash that lay in a thin layer on the ground, sniffing their fingers and tasting the burnt cinders gingerly. No one seemed willing to make a decision. The atmosphere was becoming thick with choking dust; an acrid taste of burning invaded the nostrils, mouth and throat with every breath. It was becoming unbearable. Yet, uncertainty held them on the spot.

Finally, two of the older males in the group began to walk off, heading out across the plain away from the mountain. The rest of the group followed, at first stepping cautiously into the powdery dust that covered much of the landscape. Moving steadily in their effort to escape the mountain's heavy atmosphere, the group began to spread out as little clusters slowed down by infants and juveniles became detached. Mid-morning found three of the group, two adults and a youngster, well separated from the rest, who could still be seen in the distance. They walked on, unhurriedly, the adults stepping carefully into each other's tracks, the juvenile walking to one side, occasionally drifting in to walk beside an adult and then drifting back out as youngsters will. By now they had become more accustomed to the rumblings of the mountain in the background and paid less attention to it. A light rain began to fall, dampening the ash carpet so that the little whorls of dust no longer rose under them as they stepped into it.

Then the mountain gave a violent series of explosions, spewing out great streams of lava, steam and ash. The unexpected loudness startled them, and one paused to glance back and see what was happening. Startled by the explosions, a herd of a now extinct horse species clattered across their tracks behind them. Clouds of swirling gases swept down the mountainside, obliterating everything that stood in their path. The three pressed on, a new urgency to their stride.

But they would never make it. Caught in the choking clouds of hot gas and dust spewed out by one especially violent explosion, they succumbed to the mountain's outburst. Their bodies and their tracks, the latter hardened into concrete by a rain shower that followed, were covered in ever deepening layers of ash filtering down from the sky as, day by day, the mountain vented its fury.

Nearly four million years later, one August morning in 1978, the fossil-hunter Mary Leakey would stumble across those tracks while carefully scraping away the surface layers in search of fossils at a site now known as Laetoli in northern Tanzania. With growing astonishment, she and her assistants would lift the layers of hardened volcanic tuff on a few more inches of track, only to see the footprints disappearing tantalisingly under the excavation's edge ahead of them. They would uncover some 50 metres of trackway preserved under the layers of cold ash.

We marvel at the spectacle and wonder about the unnamed beings who made those tracks so long ago, hearts beating in the choking dust clouds. All we can know is that these creatures were a part of the broad tree of our ancestry, but we will never know for sure which species they belonged to or whether they lay on the direct line of our ancestry from the apes or on a side-branch that became extinct long before our ancestors began to look all that much different from our ape cousins.

The Anatomy of a Difference

We humans are natural classifiers. Throughout history, our ancestors readily classified the species of plants and animals that they came across as they hunted and gathered in the forests and woodlands of the world's inhabited continents. These natural classifications were (and still are) based on physical similarity. Species that look alike are assumed to be related. That much is a natural inference from everyday experience: children tend to resemble their parents, whether they be human, animal or plant. Given this, it was perhaps inevitable that we humans came to see ourselves as standing somewhat apart from the other animals. To be sure, we bore unmistakable similarities to the apes and monkeys that were without question our zoological cousins. But cousinhood was perhaps about as far as it went. The differences between us and the apes remained profound. We were blessed with big brains and great technical intelligence. We had, after all, founded cities and nation-states, built temples and dams, travelled the world by canoe and ship and developed the most powerful weapons of destruction the world had ever seen; we had language and culture, wrote plays, and spoke of gods and goodness. And there were the obvious physical differences too: we walked upright while the monkeys and apes ran on all fours like the proverbial beasts of the field; we

13

were notably hairless, blessed with a coordination that allowed us to throw spears and stones with aimed precision.

These differences between ourselves and the rest of creation were, of course, reinforced in the Judaeo-Christian tradition (even if not in all religious traditions) by the belief that we humans were in some sense special in the eyes of the Almighty. We humans were blessed with souls breathed into our bodies by some unfathomable divine action. In the older biological theories of the eighteenth and early nineteenth centuries, evolution was seen as progressive, with humans at the apex immediately below the angels and, ultimately, God himself. For these early biologists, evolution was a linear affair, and we could gauge the relative ages of different species simply by comparing their degrees of complexity. Humans, being by far the most complex species, must have been around for much longer than any of the others because we had had more time in which to advance along the great chain of being between its first (presumably virus-like) stages and its final pinnacle (godhead itself).

The situation changed dramatically in 1859 when the biologist Charles Darwin published his landmark book *On the Origin of Species*. Darwin's view was radically different from those of his predecessors because he argued that evolution was not linear and progressive but, rather, more like a branching tree. Moreover, change in a species' appearance (and thus ultimately the evolution of completely new species) arose, Darwin argued, as a result of natural selection acting on inherited material that parents passed on to their offspring. As the environment changed, so species were forced to respond, adapting their form and behaviour to the new conditions. Species that failed to do so simply went extinct and their lineages were lost for ever.

One shocking implication of this new perspective was that we humans were not necessarily the pinnacle of evolution: all currently surviving species were in principle equally 'good' merely by virtue of the fact that their existence implied that they were well adapted to current (or, at least, recent) conditions. Even those life forms that are now extinct should be regarded as evolutionarily successful. So far from being an evolutionary disaster, the dinosaurs were in fact a hugely successful group: they existed for considerably longer than we humans have been on our self-appointed pedestal.

One implication of all this was more upsetting to Darwin's Victorian audience than anything else: it was the suggestion that we humans were merely a sub-set of one lineage in the grand panoply of life. The primates seemed the most likely lineage, and among the primates the apes seemed to be most like us. We were not, as we had so long believed, the product of divine special creation. We were just another ape. Our history was intimately bound up with the histories of other closely related species.

Despite the implications of these new ideas, the zoological classifiers (or taxonomists) of the nineteenth century continued to place us in our own branch of the primate tree of evolution, where we occupied a position of solitary splendour among the apes (Figure 1a, overleaf). This view held sway until a mere decade or so ago. The real importance of this traditional taxonomy, however, was that, in combination with Darwin's theory of evolution by natural selection, it implied a deep ancestry to the ape-human group. Aside from our obviously bigger brains, one key difference between us humans and our ape cousins weighed heavily on the classifiers' minds – namely the fact that we walked upright and had an anatomy to match (long, powerful legs and short, weak arms) whereas all four great apes (the two chimpanzees – the common and the bonobo – the gorilla and the orang-utan) shared a common quadrupedal style of locomotion designed principally for shin-

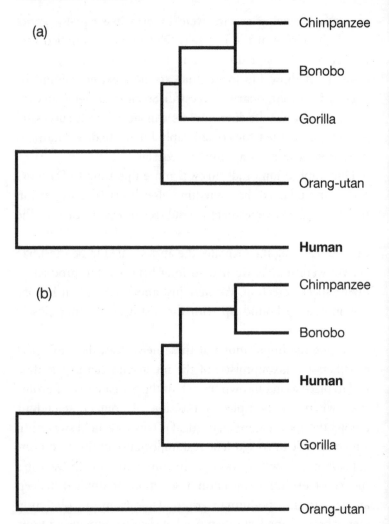

Figure 1: Our relationships with the apes. (a) The traditional view that prevailed until about 1980: the human lineage is distantly related to the other four species of great apes, from whom our ancestors separated off some 15 million years ago. (b) The new view accepted since about 1980: humans are most closely related to chimpanzees, who, together with the gorilla, form the African great ape clade which split into the three lineages some 5 to 7 million years ago. The orang-utan separated off from the ancestors of this group around 15 million years ago.

ning up tree trunks (with an associated pattern of very strong arms and short, weak legs). This view was reinforced by the successive specimens of ancestral hominids that were unearthed during the ensuing decades of the twentieth century. All, without exception, were upright walkers with long, powerful legs and relatively weak shoulders and arms. The human ancestry must indeed have been very deep.

The inevitable conclusion was that the last common ancestor of the human and great ape lineages must have predated the separation of the orang-utans from the other African great apes (the gorilla and the two chimpanzee species, the common and the bonobo or pygmy). Since the orang-utan's ancestors appear in the Asian fossil record around 15 million years ago, the facts of geological history suggested that this last common ancestor must have lived sometime prior to that date.

The picture, then, was one of an ape-like family that divided into two quite distinct lineages some 15 to 20 million years ago – one leading eventually to the four great apes as we find them today, the other via a series of now extinct species to ourselves. That the fossil record was completely devoid of any human-like fossils between this supposed separation some 15 million years ago and the point just short of four million years ago when the first identifiably human-like ancestors appear was an inconvenience, but perhaps not too great a matter for concern since there were few or no ape fossils either. The geological processes that result in the fossilisation of bones are rather haphazard, and the forested habitats in which most of these species were assumed to have lived are very poor environments for fossilising anything.

Though this might seem a rather tenuous assumption to have made, it is perhaps not unreasonable. We cannot trace the recent fossil ancestry of any of the three living African great apes. The apes as a whole have an excellent and abundant fossil record spanning the period between their first appearance some 20 million years ago until around 10 million years ago, but after that it becomes very patchy with few or no fossils outside of Asia where the orang's ancestors and close relatives (some of whom, like the aptly-named 200-kilogram *Gigantopithecus*, were among the largest primates ever to walk the earth). In contrast, today's African great apes (the gorillas and the chimpanzees) seem to appear out of nowhere. There are no species in the fossil record of the last five to ten million years that are obviously related to them.

A Genetic Spanner in the Works

This view remained unchallenged until the late 1960s when two Californian geneticists, Vince Sarich and Alan Wilson, had the temerity to suggest that the common origin between ourselves and the apes might be of much more recent date, possibly as recent as three million years ago. Their grounds for making this quite outrageous claim were the similarities in the genetic codes of humans and the African great apes. New developments in the science of genetics following on from the cracking of the genetic code in the 1950s had led to the idea that DNA – those immortal chemical strands curled up inside every living cell that carry the information for the building of new bodies – might be used as a kind of biological clock.

This suggestion arose from the recognition that the structure of the genetic code could change over time. It came about because of an imperfection in the way DNA, the molecules that make up the genetic code, copy themselves during the course of reproduction (the process we refer to as 'mutation'). In many (but not all) cases, such changes have no effect on the way the body functions, and these minor differences in the individual's genetic signature accumulate down through the generations as a kind of genetic baggage.* Since the rate at which these 'hidden' changes occurs is roughly constant across time, the number of differences between any two individuals represents a rough measure of the amount of time since they last shared a common ancestor. This contrasts, of course, with what happens when a bit of DNA codes for a part of the body: in this case, the success or otherwise with which that part of the body works will directly affect whether that particular gene gets passed on to the next generation; as a result, natural selection – Darwin's great mechanism that drives evolutionary change – can force the pace of change. Under selection, gene frequencies can change very fast – within a matter of a few generations – whereas when a gene is neutral in its effect and there is no selection pressure either way, gene frequencies change only by the very gradual accumulation of mutations.

It was this process of slow accumulation of copying errors that Sarich and Wilson thought might provide us with a kind of molecular clock that could be used to reconstruct the separation times of closely related living lineages. What upset the human evolution community was that the dates for the ape-human split suggested by the molecular analyses were far too young – around three million years ago. For those who had for decades believed that the date of the split was somewhere in the region of 15–20 million years ago, as reflected in the hard evi-

* It turns out that almost all of our genetic material is a kind of irrelevant genetic baggage from our past. The human genome (the sum total of all the DNA in our 24 pairs of chromosomes) contains the equivalent of about one billion genes, but only about 30,000 of these are involved in producing the human body and all its attributes. The rest – known in the biological trade as 'junk DNA' – seems to be a combination of structural elements and viruses that have vicariously inserted themselves into our DNA over the course of time since life first evolved. This 'junk DNA' sits quietly in its molecular den, effortlessly leapfrogging its way through evolutionary time courtesy of its host's ability to reproduce. Since most of this DNA has no effect on the body, it is not influenced by natural selection, and so is subject only to the internal processes of mutation. It is this segment of DNA that provides the basis for the molecular clock.

dence of the fossil record, this incredibly young date seemed absurd. So absurd, in fact, that they were inclined to scoff at the validity of this newfangled genetic technology. It was obviously not worth the chemicals being wasted on it if it produced results that were so out of line with what the fossil record was telling us. Either the clock simply did not work as a clock or the calibration was wrong.

Eventually, and perhaps inevitably, the geneticists turned out to be right. The original Sarich-Wilson insight was confirmed: the molecular clock really does seem to work and can be used to determine the date of divergence between two lineages. Living humans and the great apes do have a much more recent origin than anyone had ever imagined. Even though the estimates for the age of the last common ancestor between chimpanzees and humans had to be revised upwards a bit from the original value of three million years, we were talking about a much more recent date than the fossil record had suggested. The current best estimates suggest a date for the last human-ape common ancestor not later than five–seven million years ago. The more bits of chromosome that were analysed, the more the estimates tended to converge on this crucial period. In geological terms, of course, this is still barely the twinkling of an eye. Primates, after all, have been around as a family for more than 65 million years.

But the really astonishing conclusion from all this was that humans were more closely related to (i.e. shared a more recent common ancestor with) chimpanzees than either of them were to the gorilla or the orang-utan. This finding was to completely up-end our understanding of primate taxonomy and our own origins. Now, it seemed, so far from being a separate branch with a long and unique evolutionary history, the human lineage was in fact deeply embedded within the great ape family. And more than that: it was specifically within the African great ape family that humans nestled so cosily (Figure 1b, page 16). The molecular evidence unequivocally demonstrated that it was the orang-utans, not we humans, who were the outsiders at the family feast. It was the orang-utan that had had the long independent evolutionary history isolated on the Asian landmass for some 15 million years or so, far from the ancestral great ape populations in Africa where they too had once originated.

This unsettling conclusion meant both a radical rewriting of the zoological classification of the apes and a rather drastic reassessment of our own relationship with our cousins. Indeed, so close was the relationship between humans and chimpanzees (as the widely quoted claim has it, we humans share 98.5 per cent of our DNA with the chimpanzees) that it seemed we were really just chimpanzees gone mad. So far from being our cousins, the chimpanzees are – taxonomically speaking – in reality our sister species.

The Patchwork of Evolution

If humans were just highly evolved chimpanzees that shared a recent common ancestry with the living chimpanzees, how should we view our oldest known fossil ancestors, the australopithecines (or 'southern apes'), whose remains had been unearthed from many sites in eastern and southern Africa over the course of the last century? The earliest of these species roamed the African savannahs around four million years or so ago,* and the last of them became extinct as recently as 1.2 million years ago.

Conventional wisdom had always set these species off as very different from the other apes that must have occupied the

---

* Recent fossil finds near Lake Chad in West Africa and the Tugen Hills in Kenya have yielded what may turn out to be very early members of this family, dated to around six million years ago.

African forests at this time. What made the australopithecines seem so different from conventional apes was the fact that they were habitually bipedal. In other words, they walked upright on two legs just as we do. To be sure, their hips were less well adapted to the kind of striding walk characteristic of modern humans (that would not appear until the first members of our genus, *Homo erectus*, hove into view around two million years ago) and so they probably tended to shamble rather than stride when walking on the ground. But this perhaps reflected a compromise on the fact that they seem to have been much more arboreal than we now are, able to climb about in trees more skilfully than we can. Nonetheless, the one thing they could not have been was habitual walkers on all-fours like all the living great apes.

So much was abundantly clear when Mary Leakey uncovered that section of trackway buried below the surface at Laetoli in 1978. There, frozen in time and unseen for nearly four million years, was unmistakable evidence of the passing of a bipedal species. This trackway is not the unsteady stumbling of a normally quadrupedal ape walking for very short distances on two legs because the hot lava ash was burning its tender hands. This was an animal that habitually walked on two legs. The tracks show no signs of unsteadiness or hurry, but pass on their way across the plain, criss-crossed by the tracks of an extinct horse, illuminating for one brief minute in the harsh lamplight of an ancient volcano the passage of one of the earliest bipedal hominids. Here, one of them half turns to view what might have been a particularly loud explosion from the mountain or perhaps to check on the thunder of hooves of the approaching herd about to run them down in its own panic. One of the adults steps carefully in the footsteps of the other in the powdery ash, so that sometimes only two sets of prints, one large and one small from the accompanying child, can be seen; but at other points the larger set separates out to form two sets of partially overlapping prints. Though these three sets of footprints have sometimes been interpreted by the more excitable media as the ancestral family (mum, dad and youngster), the reality is that they were almost certainly part of a larger group whose other members were not all that far away.

But the important thing is that we can see the clear impression of a very humanlike footprint with its big toe close to the other toes at the end of the foot. This is not an ape's handlike foot with the thumb-like big toe at the back of the foot near the heel. It is very human-like. Our long foot, with the big toe joined with the others, allows very efficient striding because it provides a platform rather like a coiled spring that gives us an added push as we walk. The evidence of the trackway suggests that bipedalism predates by at least several million years the very earliest beginnings of that dramatic increase in brain size and tool use that was to lead ultimately to modern humans. This in itself was something of a surprise, since older theories of human origins tended to see our large brain, bipedal stance, hunting and technological skills as all being part of the same adaptive complex. We began to walk upright, so the conventional view supposed, because it allowed us to throw more effectively the spears and stone missiles that our big brains had developed to enable us to hunt. In fact, the evidence suggests that all these traits evolved piecemeal over a period of several million years.

By far the earliest of these traits to evolve was bipedalism. Aside from the evidence of the Laetoli footprints, there was the evidence from the size and shape of the pelvis and leg bones of the earliest fossil australopithecines who lived only a few hundred thousand years after the Laetoli individuals. The apes share with the quadrupedal monkeys a pelvis that is long and thin, designed to provide secure anchor points for hind limbs built to climb trees and run quadrupedally. The long shape in turn provides a beam off which to hang the gut. In contrast, the pelvis of modern humans, and that of all our fossil ancestors back to the earliest australopithecines, is bowl-shaped. It is designed to provide both a stable platform on which to balance the trunk and a bucket-like support for the guts which would otherwise flop forwards uncontrollably and hang around our knees (as is so painfully obvious in the case of those of us whose guts have come to exceed the capacity of our particular buckets). These are features closely associated with a bipedal gait. The long thin pelvis of the monkeys and apes does not provide a sufficiently stable base for our heavy trunk. Moreover, the section of the pelvis that juts out behind the legs obstructs their movement when the body is raised upright. In the bowl-shaped pelvis of modern humans, the hip joints are set wider apart and there is nothing to obstruct the swing of the thighs during striding.

The apes as a group are also characterised by rather short stubby thighs that allow the mass of the body to be brought close to the trunk of the tree when climbing. In effect, apes sit on their haunches when climbing, alternating between pulling themselves vertically upwards with their strong shoulder muscles and then bracing the feet against the tree trunk while reaching upwards for the next cycle. It is a very efficient design that allows then to shin up vertical tree trunks with remarkable speed. We can do this too, but we are not particularly skilled at it because our long thigh bones force our centre of gravity to lie too far out from the tree trunk and we cannot bring enough pressure to bear on our feet to allow them to grip the trunk well enough to support our weight. (We have to use a loop of rope around the feet to provide them with enough grip on the vertical trunk.) The first of these problems is all too apparent when you try to squat with your feet flat on the floor. You'll find that your long thighs set your body too far back behind the feet, and you topple backwards. Or your shins quickly start to ache from the strain of using the muscles to try to hold the body in position above your feet. Apes, with their shorter thighs, have no problem with this exercise.

Australopithecines had longer thighs and shorter arms than is typical of the apes, indicating that they were more used to travelling bipedally even if they were not as efficient at this as the later hominids. One likely reason for this lesser efficiency is that the australopithecines were still partially arboreal, so they could make the most of both the open savannahs and the trees in the woodlands that bordered them. In contrast, their sister-species that were later to give rise to the chimpanzees and gorillas remained well ensconced in the forests and had little need to travel extensively on the ground.

It is interesting that the bonobo – perhaps the most human-like of all the great apes – can occasionally be seen striding for quite long distances across the forest floor on two legs, sometimes carrying small branches in one hand. They look so uncannily human when they do this that it is hard not to believe that we are watching a group of australopithecines. Indeed, one of the features that makes the bonobos look more human is that they have slightly longer legs than is typical of the other two African great apes – their sister-species, the common chimpanzee and the gorilla: it makes them look more gracile, giving them the impression of being smaller than the common chimpanzees, who seem much more solid and squat by comparison. One other key feature in which bonobos differ from the other apes is in their ability to lock the knee. While the other apes when they walk bipedally do so with bent knees, bonobos are able to straighten their legs, and this allows them to remain bipedal for longer than the other apes.

But true striding bipedalism like that characteristic of modern humans was still a long way off, appearing for the first time in the fossil record only with the earliest members of our own genus, the species *Homo erectus*, around two million years ago. This required a further tweaking of the anatomical frame. In humans, the thigh bones angle inwards so that the knees meet. Ape thigh bones are placed vertically below their point of attachment to the hip, so that when they walk bipedally they are forced to sway from side to side, rather as a sailor does when he first steps ashore. In the ape's case, however, the waddling style of walking is due to the fact that the legs and feet are placed out to one side of the body's midline, so at each step the animal is forced to sway to the side in order to bring its centre of gravity above its foot in order to prevent itself falling over. Our angled thighs mean that our feet are side by side rather than being spaced a hip's width apart so that, when we walk, it requires only a gently graceful sway to ensure that our centre of gravity remains balanced above each foot as the other is lifted to step forwards.

This rather unusual design allows us to walk bipedally for very long distances without putting too much strain on the legs and stomach muscles that support the body's weight when we are upright. This small but significant change in anatomical design probably coincided with a shift in foraging style from relatively short journeys around a modest-sized home territory in the woodlands that bordered the forests to a more nomadic lifestyle based on long migrations between foraging areas. This shift in lifestyle seems to coincide with a period of dramatic climatic instability around two million years ago that resulted in a cooler drier climate in Africa. The resulting loss of forest and corresponding expansion in the grasslands and woodlands must have placed increasing pressure on the ape populations in the forests themselves and the australopithecines that occupied the adjoining woodlands. It seems that, under pressure from this climatic stress, some australopithecines took advantage of their partial adaptation to bipedal walking to exploit the open habitats even more effectively than they were already doing. In contrast, all the other great ape species retreated deeper into the forests, their geographical ranges contracting as the forests contracted.

Our unique bowl-shaped pelvis provides us with a painful reminder of the fact that evolution is something of a piecemeal process. For most primates, the business of giving birth is relatively quick and trouble-free. For humans, however, giving birth is, to put it literally, a labour. This is because we are trying to squeeze a baby with an unusually large head (for a primate) through a hole (the birth canal through the pelvis) that is, relatively speaking, unusually small (for a primate of our size). This is the unfortunate result of the fact that, when our pelvis rounded itself out to act as a base for the torso and head, the bones that surround the birth canal (the hole through the front of the pelvis through which the baby passes during birth) were forced towards each other. This was not too much of a problem at the time because australopithecine babies' heads were not all that much bigger than chimpanzee babies' heads. So far, so good – at least for a couple of million years. But once the human brain started its rapid increase some time from around half a million years ago, the problem became more acute. By then, however, we were fully committed to bipedalism. Widening the pelvis would have seriously disadvantaged women's manoeuvrability: they would have waddled rather than walked or run, and so would have been easy prey for predators.

Instead, our ancestors opted for an alternative solution to the dilemma: they reduced the length of pregnancy. In all mammals (and especially primates) except humans, the length of pregnancy is determined by the size of the species' brain. Birth is the point at which the baby's brain more or less reaches its full adult size, and there is relatively little growth after birth. If humans were to have the gestation period appropriate for the size of their adult brain like other mammals, pregnancy would last an eye-watering 21 months. To cut through the Gordian knot, our ancestors opted to give birth at the earliest point that the infant could survive outside the womb and finish off brain growth afterwards. Even so, it is a very tight squeeze, which is why giving birth in humans is a labour while it is not for our immediate cousins. In fact, the squeeze is so tight that the ligaments between the two halves of the mother's pubic bone become more elastic during late pregnancy, allowing the two halves of the pelvis to separate slightly during birth, thus enlarging the hole through which the baby has to pass (which is why many women complain that they cannot get back into their old trousers after they have produced their first baby). At the same time, the plates of the baby's skull (which, unlike those of adults, are not yet welded together) slide over each other at the edges as the head forces its way through the birth canal, so helping to make the skull just that little bit smaller.

This, then, is why human babies are so helpless for so long after birth: human babies do not achieve the same state of brain and body development that an ape baby has at its birth until they are about a year old. And that, in turn, is why human babies that are born prematurely are such a cause for worry and concern. Human babies born at the end of a normal pregnancy are only just capable of survival; those born any earlier are really cutting the thread of life very fine indeed.

Neither the australopithecines' brain size nor their way of life seems to have been radically different from those of their chimpanzee cousins. Their brain size is well within the chimpanzee/gorilla range. The real spurt to brain-size evolution in the hominid lineage occurs much later. With the appearance of the first members of the family *Homo* around two million years ago, we see the beginnings of a rapidly accelerating increase in brain size. Even so, it is not until the appearance of our own species, *Homo sapiens*, around half a million years ago, that this acceleration really begins to take off. Brain size rises exponentially through time as we pass from the late populations of *Homo erectus* (the last of the pre-human species) through the early (or archaic) *Homo sapiens* populations towards ourselves. Though it has to be said that it is not with us that brain size reaches its apogee: ironically, it is among the much maligned and now extinct Neanderthals that brain size achieves its greatest volume (Figure 2).

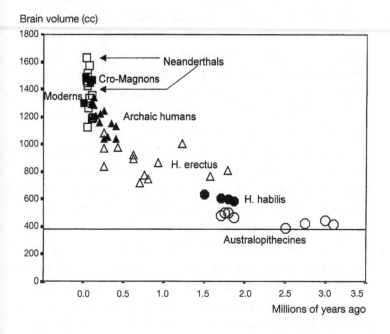

Figure 2: The pattern of brain-size evolution in the hominid lineage. The estimated brain volume of individual populations of hominid fossils is plotted against the time when they lived. The early australopithecines had brains well within the living great ape range (the horizontal line shows the brain size of chimpanzees). The largest brains of all were those of the Neanderthals. (Source: Aiello and Dunbar, 1993.)

Meanwhile, the archaeological record seems to proceed along an even more disconnected track. The stone tools associated with australopithecine remains are pretty unimpressive, consisting mainly of weakly fashioned stones – so weakly fashioned in many cases that it is far from clear whether these are deliberately prepared tools or accidentally broken rocks picked up for a specific purpose of the moment and abandoned thereafter. They are not that different from the stone tools used today by the chimpanzees of the Taï forest in Guinea, West Africa, who employ stones as hammers for cracking open the hard shells of palm nuts. These kinds of stone tools reflect a way of life that is based largely on the gathering of vegetable foods in the time-honoured traditions of all monkeys and apes. The difference seems only to be that at least some of the australopithecines were doing this in more open country habitats where succulent fruits were less common and more use was made of hard nut-like fruits and the more fibrous underground roots of plants adapted to highly seasonal habitats.

Not until the appearance of *Homo erectus* around two million years ago is there any evidence for a dramatic improvement in the range and quality of tools. Now we have more deliberately shaped handaxes. But after that, stone tools remain all but unchanged for the better part of two million years, despite the fact that the brain doubles in size over this same time period. The really big change in stone tool production only occurs as recently as 50,000 years ago with the so-called Upper Palaeolithic Revolution – around 100,000 years or so after the human brain reaches its modern size. The change is very sudden (hence the reason it is usually described as a revolution). Before this point, we have rather crude but functional tools; afterwards, there is a profusion of more delicately constructed implements (knifelike blades, borers, arrowheads) as well as items designed to serve functions other than mere food-extraction: by 20,000 years ago, we are into awls and needles, brooches and Venus figurines.

In sum, then, those traits that have long been considered to be the key markers of humanity (bipedalism, big brains and tools) seem to derive from quite different time periods. So far from appearing as an all-in-one suite of characters spun out of one massive super-mutation, they appear in dribs and drabs across the whole range of our evolutionary history, beginning more or less as soon as our ancestors parted company with the other apes. There is no one point in our history at which we can safely point and say: 'Ah, and now we became human!' The point at which we make this claim depends entirely on which of the many marker traits we want to take as being the one that makes us 'especially human'. Perhaps we might be better advised to see the history of our species as one of increasing degrees of humanity which only finally came together as a unique suite a mere 50,000 years ago with the Upper Palaeolithic Revolution.

The Quirks of History

And yet we modern humans *are* very different from our ape sister-species. How can this be?

The answer, as it turns out, is relatively simple: in one important sense, the difference is more apparent than real – an illusion created by the fact that modern humans are a *very* recently evolved species. To be sure, some of our traits, like bipedalism, are ancient and probably reflect a period of intense selection for a two-legged style of locomotion very early in our history. But we ourselves as we are now are a very recent offshoot from this relatively ancient lineage.

Homo erectus was the first of the true members of our genus, the genus *Homo*, appearing for the first time around two million years ago. In the subsequent million years, it spread all around the Old World from its African homeland, as far east as north-eastern China, into what was eventually to become the islands of the Indonesian archipelago, and north into Europe. It holds the record for the longest surviving hominid species, having survived until well past half a million years ago. The long period of its existence was marked by a degree of stability in the species' general form and style that lasted for the better part of one and a half million years. Some changes are, of course, to be expected over so long an interval – there was, for example, a gradual increase in brain size with time – but by and large what we see is minor tinkering on a theme rather than anything substantively new.

Then some time around half a million years ago, one of the African populations of *Homo erectus* began to undergo rapid evolution towards larger brain size and lighter body build. In a relatively short space of time, it had spread across the continent and into the Near East and Europe, replacing the populations of *H. erectus* that had previously lived there. The species *Homo sapiens* had arrived, though it still exhibited many primitive *erectus*-like features, including physical robustness, heavy brow ridges and a still slightly less than modern brain volume. To distinguish these early humans from ourselves, they are usually referred to as *archaic Homo sapiens* or even given a separate species name, *Homo heidelbergensis* (named after the site near the German city of Heidelberg where the first specimen was found). Meanwhile, *Homo erectus* continued to survive in Asia where it remained free of competition from the new species, perhaps even surviving until as late as 60,000 years ago when the first modern humans swept in from the west.

Meanwhile, in Africa, the new species was undergoing yet another phase of rapid evolution. Around 200,000 years ago, a lighter and even more gracile variant on the human theme began to evolve somewhere in (possibly eastern) Africa and eventually replaced the older heavy-bodied populations of archaic humans. Known generically as 'anatomically modern humans' (or 'AMH'), they spread with remarkable speed: by 150,000 years ago they had probably replaced all the older archaic human populations throughout Africa, crossing the Africa–Eurasia landbridge to enter the Levant (that eternal crossroads between Africa, Europe and Asia) around 70,000 years ago. From there, these mobile, highly organised hunters raced across the southern Asian landmass, crossing the waterways that separate Asia from Australia by 60,000 years ago, and backtracking into Europe by 40,000 years ago. By 15,000 years ago, they had trickled across the Bering Strait, separating Asia from North America, when a period of low sea level exposed a landbridge across the strait. By 12,000 years ago, they had surged down the long American landmass to colonise the forests of the Amazon and the pampas of Patagonia, helping to wipe out the unique and remarkable giant animals of North and South America as they did so.

This explosive expansion across the globe must have happened very quickly because analysis of the mitochondrial and nuclear DNA* of a wide variety of living humans suggests that all the five billion or so humans alive today descend from a small group of about 5000 female (and roughly the same number of male) ancestors who lived around 150,000–200,000 years ago. This ancestral group must have lived in Africa, because Africa contains many more variants of modern human DNA

* The DNA that makes us what we are is contained in 23 pairs of chromosomes that lie within the nucleus of each cell; with the exception of the very small Y-chromosome (which is inherited by males only from their fathers), this represents material inherited equally from each parent. However, we also possess small quantities of DNA in mitochondria, the tiny power houses that float in the cellular fluid outside each cell's nucleus. In both sexes, mitochondria are inherited only from the mother; mitochondrial DNA thus provides a record of each individual's maternal lineage that is uncontaminated by the complications of sexual reproduction.

than the rest of the world put together. All the non-African races (Europeans, Asians, native Australians and Americans) plus a handful of scattered groups living along the southern margins of the Sahara are much more alike in terms of their DNA than are the rest of the Africans. We Eurasian-Austro-Americans are a subset of the range of variation found in Africa, and our common ancestor – whose descendants spread throughout the rest of the world – lived in Africa until a mere 70,000 or so years ago, presumably somewhere in its north-eastern corner.

It is this extraordinarily recent ancestry for modern humans that helps to explain why we seem to be so different from the other apes. The living African apes are the product of some seven million years of evolution; even the two species of chimpanzee separated some two million years ago. In stark contrast, all the humans alive today derive from an ancestor that lived a mere 200,000 years ago. This explains why – minor superficial differences like skin colour and body proportions apart – we exhibit so few real differences amongst us. We are but the babies in the great ape family tree, the new kids on the block. The fact that all the species that lie between us and our ape cousins have died out merely serves to exaggerate the differences between us. Were the Neanderthals or the later *Homo erectus* still alive – as the former were until around 28,000 years ago – the gap between humans and the other great apes might be less glaring.

The Final Mystery

It would not be right to end this part of our story without saying something about the Neanderthals, one of the most enduring and evocative of all archaeological puzzles. This extraordinarily successful race of humans inhabited Europe from the Iberian peninsula in the west as far east as Uzbekistan and Iran in west-central Asia for the better part of 300,000 years – rather longer than we modern humans have so far been in existence. Indeed, the Neanderthal fossil record is one of the best we have: we have the fossil remains of more than 270 individuals from some seventy sites. Neanderthals were clearly far from rare. Yet, suddenly within a very short space of time, about 30,000 years ago, they simply faded out. The fact that this seems to have coincided with the arrival of anatomically modern humans (a group of people known as the Cro-Magnons) in Europe from Africa some 40,000 years ago has always seemed . . . well, suspicious.

But who were the Neanderthals and what were they doing in Europe?

The Neanderthals were physically quite distinct from modern humans. In contrast to our rather gracile modern human physique, Neanderthals shared with the archaic humans that lie at the root of the human family tree around 500,000 years ago a rather heavy build. In the later Neanderthals, in particular, this is reflected in rather short heavily muscled limbs, a protruding but chinless face with a very large nose, heavy brow ridges above the eyes, an elongated low-vaulted skull leading to a distinctive 'bun' at the back of the head and a characteristic barrel-shaped chest. Although these features are enough to make most Neanderthal fossils unmistakable, a Neanderthal in clothes would probably attract only passing attention in a modern urban environment. We are sufficiently familiar with individuals who are barrel-chested and stockily built not to pay too much attention to a Neanderthal's very distinctive skull shape under a mop of hair.

This underlying similarity in physical form has perhaps been the reason why the relationship between the Neanderthals and our own ancestors, the Cro-Magnon peoples, has been something of a moveable feast. When the first Neanderthal fossil was discovered in a cave in the Neander Valley, near Düsseldorf in Germany, in 1856, the remains were at first thought to be those of a degenerate human because of the deeply bowed leg bones (assumed to be the consequence of rickets) and heavy build. However, as more and more specimens began to turn up all over western Europe and the Levant, it became apparent that Neanderthals were a widespread race. They were then assumed to be the direct ancestors of modern Europeans, giving rise to the Cro-Magnon peoples that followed them. Over ensuing years, the alternate claims that the Neanderthals were an extinct side-branch and that they were direct ancestors of modern Europeans fell in and out of favour on what was very nearly a decade-long cycle. It was, in the end, the molecular genetic evidence that finally clinched the matter and put paid, once and for all, to this intellectual rollercoaster.

The traditional view had been that the different races of modern humans had evolved from local populations of *Homo erectus*, with the Neanderthals being an intermediate step in that sequence in Europe. The molecular evidence made this untenable because it implied that all the races of modern humans had evolved long *after* the populations of archaic *Homo sapiens* had dispersed out of Africa, never mind the much earlier dispersals of *Homo erectus* populations into the farthest corners of Eurasia.

But if all modern humans were simply Africans under the skin, where did this leave the Neanderthals? The answer lay in their DNA, but – *Jurassic Park* notwithstanding – the impossibility of extracting DNA from fossilised bone made the crucial analysis tantalisingly difficult. However, since Neanderthal fossils are not very old, it is sometimes still possible to find actual fragments of cartilage or bone untouched by the fossilisation process. So it was that, during the 1990s, the Finnish geneticist Svante Pääbo and his colleagues were able to extract a few cells from the upper arm bone of the original Neander specimen. It turned out that this Neanderthal's DNA lies well outside the range of variation seen in modern humans. They could not possibly have been the ancestors of modern Europeans, though comparison with chimpanzee DNA demonstrates that they were clearly on the human tree. In fact, the differences between Neanderthal and modern human DNA suggested that they last shared a common ancestor some time around 500,000 years or so ago, placing that common ancestor close to the ancestral root of the *Homo sapiens* family tree. Similar analyses have now been done with some of the other late Neanderthal specimens, but always with the same result: Neanderthal genes were different enough to place them firmly outside the lineage of all modern humans.

The current view is that Neanderthals represent the descendents of an early migration out of Africa into Europe by archaic *H. sapiens*. There are many fossil specimens belonging to archaic humans from the period between 500,000 and 300,000 years ago in Europe. Anatomically, these individuals differ little from the archaic humans found throughout much of sub-Saharan Africa at around the same time. Since many also show some Neanderthal-like features (a rather heavy build, thickened eyebrow ridges), the suggestion that the Neanderthals represent the descendants of these archaic human populations that evolved along their own trajectory in Europe seems reasonable.

In the end, of course, the later Neanderthals developed a suite of very distinctive anatomical features. Some of these features were doubtless accidents of genetic drift as their populations evolved over the ensuing millennia out of contact with the descendants of the archaic humans in Africa. Other features (notably their short limbs, squat physique and, possibly, their large noses) may represent adaptations to the cold of the European Ice Age climates. Modern humans who live in cold environments, such as the Eskimos (or Inuit, as they are now more properly called), show similar body proportions – modern human races that have inhabited high latitude habitats for even just a few tens of thousands of years contrast strikingly with the tall slender long-limbed body builds of peoples whose ancestors colonised the hot open habitats of the tropics. Short limbs designed to reduce heat loss are a characteristic of most high-latitude mammals.

Yet, in the end, the Neanderthals went extinct. It is sobering to think that a mere 28,000 years ago – barely a thousand generations back in time – our ancestors were bumping into Neanderthal bands in the European landscape. That is almost close enough for us to be able to reach out and touch them. What caused so successful a species to die out so suddenly?

The close association between the arrival in Europe of our own ancestors, the Cro-Magnons, and the disappearance of the Neanderthals has always raised the hoary spectre of racial extermination. After all, we have ourselves been responsible for similar extinctions within living memory: the near extinction of the Australian Aboriginals and the North and South American Indians by European invaders are painful recent reminders. In some cases, those extinctions have been deliberate, as in the case of the native Tasmanians and the Cape Hottentots who were hunted to extinction by European colonists. But in other cases the demise of native populations was an accidental by-product of the arrival of the European immigrants. The devastating impact on South American Indian tribes during the course of this century of what, for immigrants and missionaries from the Old World, were trivial childhood diseases like measles has been documented by Jared Diamond in his book *Guns, Germs and Steel*. Disease remains a distinct possibility as an explanation for the disappearance of the Neanderthals, especially given that the hot humid tropics of Africa from which the Cro-Magnons had originally come are a veritable breeding ground for disease.

It seems likely that climate played a role too. Detailed models of the climate during the critical period between 100,000 and 30,000 years ago suggests that the northward extension of Neanderthal sites was limited by winter temperatures. Despite their physical adaptations to cold conditions, it seems that the Neanderthals could not cope with winter temperatures below a critical level. As the last Ice Age bit deeply into Europe, so Neanderthal living sites moved progressively southwards towards warmer conditions in the Iberian and Italian peninsulas. The evidence from ice cores tells us that global temperatures during this period fluctuated wildly from one decade to another, so it is possible that the Neanderthals found themselves increasingly caught out in the wrong place at the wrong time. The resulting attrition would have put any species under very considerable demographic pressure and made it difficult for it to recover.

In contrast, the Cro-Magnons did not seem to be quite so restricted, but are found living much further north during this same period. Despite their apparent lack of *physical* adaptation to cold climates (they were, after all, Africans and had only recently emerged from the tropics with essentially African body builds and physiology), they were nonetheless able to cope much better with the cold. This suggests perhaps a cultural difference of some sort, the most obvious being the use of clothing, since the use of both cave shelters and fire were well known to the Neanderthals.*

Although the Neanderthals have figured prominently in our folk mythology over the last century or so, in the end they seem to have been a minor European digression in the grand story of

* The controlled use of fire seems to date back to late *Homo erectus* populations, and so would have been part of the common heritage of both the Cro-Magnons and the Neanderthals.

human evolution. Events elsewhere in Asia and Africa proceeded rather differently; in Asia perhaps because there was no intermediate step between *Homo erectus* and the anatomically modern humans. The arrival of the latter from Africa some 60,000 years ago in what seems to have been an extraordinarily fast race across the expanse of southern Asia may have been the last occasion on which two genuinely different species of our family encountered each other. If they did meet, the *H. erectus* populations of China and south-east Asia did not survive the experience.

It is sobering to remember just how strange these times we live in actually are: the 28,000-year period since the Neanderthals died out is unique in the five-million year history of the human lineage in that there has been only one living species of hominid during it. Hitherto, there has probably been no time period when there have not been at least two (and sometimes as many as five) species of hominid wandering the byways of the world at the same time – bumping warily into each other from time to time. This was brought into stark relief in 2004 with the discovery of a new dwarf hominid, *Homo floresiensis*, on the island of Flores in eastern Indonesia. This tiny three-foot-high descendant of *Homo erectus* was still alive as recently as 18,000 years ago and may account for modern folk tales from the area that tell of little forest people.

The oddity of recent times has tended to exaggerate our apparent uniqueness and has perhaps been responsible for giving us a false sense of our own importance. Like all single children born late in their parents' lives, we humans have proved more than just a handful to our ageing relatives. We invariably assume that we deserve special attention.

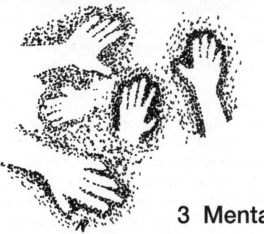

3 Mental Magic

Emerging from the dark passages of his cave, the artist blinks in the spring sunlight. He pauses to allow his eyes to adjust once more to the daylight. Then he picks up a flint-headed spear that was leaning against the wall of the cave mouth and hefts it lightly in one hand, testing the balance of the weapon as he does every time he picks it up. It was made for him by his father, a great craftsman, many years ago, and he values it above all other weapons: it has saved his life on more than one occasion when he was threatened by a cave bear or a wolf. Now, spear in hand, he begins to make his way down through the beech trees on the valley's upper slopes towards the riverbed below.

An hour's walk down river, he emerges into a clearing on the riverbank. He pauses briefly to take in the scene. Nearby, an older man sits whittling away at a length of ash stem, occasionally pausing to hold the stem over a fire until the wood darkens and hardens. The artist nods in his direction. The man smiles fleetingly in greeting. Across the clearing, two women sit together working at a deerskin spread out on the ground between them. They scrape at the skin with stone blades, carefully removing the fat and gristle adhering to the skin. It is hard and meticulous work. They cannot afford to waste good skins by making a poor job of it. Leave globules of fat and the skin will rot; but cut too deep with the blade,

41

and the skin will soon tear when used as a cape or skirt. Good skins are not so common that one can afford to be profligate with them.

A few yards from the women sit two young children. Their attention is focused intently on a bundle of old skin. One holds the tiny bundle as tenderly as a mother would a newborn babe; the other gently offers the corner of an old piece of skin to the top of a twig that peeps out from the covering, making sucking noises as she does so. She dips the skin corner carefully into a wooden bowl containing water, and offers it again to the inert form in the bundle. The children cluck tenderly in encouragement, peering attentively at their charge. The liquid dribbles off the surface of the wood as the girl squeezes the skin to drip water into what should have been its mouth. She giggles and draws back. Her companion laughs, and dabs at the wood with the corner of the covering.

The artist watches them, leaning on his spear. The younger one is his daughter, the older one her cousin. He marvels at their play, reflecting on the fact that these delicate bundles of flesh and bone that mewl and puke with such enthusiasm after women produce them should grow so rapidly into tiny imitations of an adult. It seems not so many springs since his daughter came to them, each spring marked by his annual visits to the cave where he painted the world of his imagination. Five . . . maybe six springs? It did not matter . . . what had counted for him was the miraculous appearance of this tiny doll-like bundle from his woman. He had watched its steady maturity from infant to toddler to child. Now, she seems like a small version of her mother, pretending in her turn to be mother to an inert stick of wood.

In another eight or nine springs, he would find her a mate. Some strong young man from another band. And together, they would give him grandchildren to dandle on his knee, to give him joy in his old age. But – he sighs inwardly – the way ahead was fraught with uncertainties and dangers. Who knows what trials *and tribulations are before her, what roots lie in her path to make her stumble, what dangers lurk in the shadowy woods through which life winds its uncertain way? A pucker ripples across his brow at the thought of what might befall her in the years ahead.*

Here is a very simple test you can do with any child. Sally and Ann are two dolls. Sally has a ball. She puts the ball under the cushion on the chair. Then, she leaves the room. While she is out of the room, Ann takes the ball out from under the cushion and hides it in the toy box on the other side of the room. Later, Sally comes back into the room. Where does Sally think her ball is?

Up to the age of about four years, a child will instinctively say: 'Sally thinks the ball is in the toy box.' A child of this age cannot distinguish between its own knowledge of the world and that of other individuals. But between the ages of four and four and a half years, the child passes through a rapid phase of understanding. From about four and a half onwards, it will answer the question by saying that Sally thinks her ball is under the cushion, '. . . but I know it's not.' At that point, the child is able to recognise that another individual can have a belief about the world that is different from its own, a belief that it knows (or at least thinks) is untrue. At this point, a child is said to have acquired a 'theory of mind' – it instinctively understands that others have minds of their own not unlike what it experiences as its own mind. This form of understanding, sometimes also known as 'mind-reading' or 'mentalising', is a remarkable and crucial feature of human psychology.

Tests like the Sally-Ann test are called 'false belief' tasks because to pass them the child needs to understand that another individual can hold a false belief (one the child knows to be untrue, or at least one that is different from the one the child supposes to be true). There are now a number of these tests. Another is called the 'Smartie Test'. In this test, you show the

43

child a tube of Smarties and ask: What do you think is in the tube? The inevitable answer is: Smarties. You take the cap off and the child sees that the tube actually contains pencils. After replacing the cap, you then say to the child: 'I'm just going to bring your friend Jim into the room. What do you think Jim will say is in the tube?' Up to the age of four, children will invariably say 'pencils' because they cannot distinguish between their own knowledge of the situation and someone else's; but after about four and a half, they will reply, with rapidly increasing conviction, 'Smarties.'

Tests of this kind have become benchmarks of children's developing abilities for inferring the mental states of others. They represent a critical Rubicon in the process of child development because they demarcate the moment at which children can begin to engage with an imaginary world that is not physically present. They can now begin to engage in those forms of pretend play that are so characteristic of younger children – to *imagine* that the doll really is alive and can suck liquid from the end of a rag or from a baby bottle. They can take part in dolls' tea parties, pretending that the empty teapot really does contain tea to pour into the cups, and afterwards that they can themselves drink 'real' tea from the patently empty cups.

Herein lies the great mystery of child development, for children are not born with this ability. Infants and toddlers alike treat the world as being exactly as they experience it. They cannot imagine that it could be other than what they perceive it to be. They lack the ability to imagine. And because they cannot imagine that the world is other than what they know it to be, they cannot suppose that another individual – child or adult – believes something to be the case that they *know* is not. And as a result, they cannot do something that is in some ways the hallmark of the grown-up world: they cannot exploit another individual's view of the world to feed them a lie.

The Art of Mind-reading

At this point, I need to introduce a technical term. Some decades ago, philosophers interested in the nature of minds coined the term 'intentionality' to refer to the kinds of mental states that we have when we are conscious of holding some kind of belief, desire or intention. The term refers collectively to mind-states like knowing, believing, thinking, wanting, desiring, hoping, intending, etc. It refers to the state of being aware of the contents of your own mind. Intentionality can be conceived of as a hierarchically organised series of belief-states. In this scheme of things, computers are zero-order intentional entities: they are not aware of the contents of their 'minds'. Some living organisms such as bacteria (and perhaps some insects) may also be zero-order intentional beings. Most organisms that have brains of some kind are probably aware of the contents of their minds: they 'know' that they are hungry or 'believe' that there is a predator under that bush over there. Such organisms are said to possess first-order intentionality. Having a belief about someone else's beliefs (or intentions) constitutes second order intentionality, the criterion for theory of mind (or, as it is more often known in the technical literature, ToM). Jane *believes* that Sally *thinks* her ball is under the cushion. Jane has two belief states in mind (her own and Sally's), so theory of mind is equivalent to second-order intentionality.

We humans can clearly go beyond this level, however. Figure 3 (overleaf) illustrates an everyday example of adults engaged at successive intentional levels: in this particular example, *Wife* is working at first-order intentionality, *Stranger* at second-order and *Husband* at third. Equivalently, Peter *wants* Jane to *suppose* that Sally *thinks* that her ball is still under the cushion. Sally is in first-order intentionality, Jane in second and Peter in third. It

Figure 3: Intentionality for beginners. While at a party, husband sees wife deep in conversation with a stranger. Wife is in a first-order state of intentionality (she knows what she thinks); stranger is in a second-order state (he thinks that the wife believes something); while husband believes that stranger has a false belief about wife's intentions. Stranger demonstrates evidence of the state known formally as 'theory of mind' (or mind-reading), because he has a false belief about someone else's state of mind (something that is possible only with second-order intentionality). (Drawing by Arran Dunbar; reproduced with permission of the publishers from Barrett et al, 2002.)

seems that there is an upper limit on what we can do in this respect. Conventional wisdom suggests that adult humans have an absolute upper limit on the levels of intentionality that they can cope with at about five or six orders:

Peter *believes* [1] that Jane *thinks* [2] that Sally *wants* [3] Peter to *suppose* [4] that Jane *intends* [5] Sally to *believe* [6] that her ball is under the cushion.

If your mind has just been turned inside-out by this sentence, it is not too surprising: few adult humans can keep who's thinking what here straight because, in the limit, there are too many orders of intentionality (marked out by the numbered verbs in italics) for us to keep track of. Most everyday situations probably require no more than second-order intentionality and, in actual practice, the limit for most people is probably about fourth or fifth order: Jane thinks [1] that Sally *wants* [2] Peter to *suppose* [3] that Jane *intends* [4] Sally to *think* [5] that [something concrete is the case]. We've just lost Peter *believing*.

We know this is where the limit is because we ran some tests to see just how well normal adult humans do. At the time we ran the tests, no one really knew the limits to human abilities in this domain. All the work had focused on theory of mind (second-order intentionality) and most of it had been carried out on young children around the transition point where they acquire ToM. In order to explore the capabilities of normal adults, we devised some ToM stories that went up to six orders of intentionality.

I would like to have been able to say that we set sixth-order intentionality as the limit on these stories for some incredibly sophisticated scientific reason. Alas, in reality, the truth is that I found it impossible to write a convincing seventh-order ToM story . . . The stories were short vignettes of everyday life in about 200 words: someone wanting to make a date with a girl that he thought fancied someone else, or someone wanting to persuade her boss to give her a pay rise by pretending that she had been offered another job elsewhere. Anything longer than sixth order proved so tortuous that I ended up getting completely confused myself. A bottle of whisky later, and well into the early hours of the morning, I gave up and settled for sixth-order stories.

We gave the tests to about 120 university students. They were read the stories and were then asked to answer a series of questions about who was thinking what. Between 80 and 90 per cent of the subjects answered the questions correctly at any given level up to fifth-order intentionality. That seemed very reasonable. But the subjects' performance went decidedly pear-shaped with sixth-order questions: only about 40 per cent of them got these right, as shown in Figure 4. This is a dramatic and sudden collapse in what had, up to that point, been a very competent performance by a large sample of young adults with above-average IQs. These results seem to be quite robust, since we were able to confirm them in a second study carried out some time later by Jamie Stiller using stories that went up to ninth order.

We know this sudden collapse at around fifth order is not just a memory problem, because we also asked them questions about the factual content of the stories in between the various ToM questions, and they had no problem at all with these. They

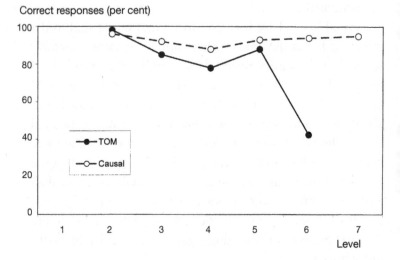

Correct responses (per cent)

Level

Figure 4: Humans seem to be limited to about five orders of intentionality. When presented with short stories about a set of people's actions and beliefs (solid symbols – 'TOM' stories), most subjects can make correct inferences about fifth-order intentional states (A believes that B thinks that C wants D to suppose that E imagines . . .), but only a minority can cope with sixth-order. That this has something in particular to do with mental states is suggested by the fact that subjects can cope easily with causal reasoning tasks up to seventh-order (A causes B, which influences C, which triggers D, which causes E, which drives F, which causes G, which finally results in H: open symbols – 'causal' stories. (Redrawn from Kinderman et al, 1998.)

remembered the main factual events in each story. We also gave them a simple factual story detailing a sequence of causally related events about an old man setting fire to himself when he fell asleep smoking. This story shared the same kind of embedded hierarchical structure as the ToM stories (A causes B, which causes C, etc). Subjects had no problems with this story: the percentage of correct responses remained constant at about 90–95 per cent right up to seventh-order concatenations of causal sequences ('When A happened, B followed, which set C off, which resulted in D, which triggered event E, which gave rise to F, which precipitated G'). So the problem seems to be something to do with the reflexivity of mental states rather than just causal sequences *per se*.

It is important to remember that not all human adults can aspire even to these dizzy heights. There is considerable interest in the possibility that schizophrenic individuals (and perhaps even those suffering from bipolar depressive disorder) score poorly on theory-of-mind tasks: at least during the more extreme phases of their conditions, it seems that they may not be able to pass second-order theory-of-mind tasks. In these cases, the ToM deficit may be acute rather than chronic – that is to say, inability to perform on ToM-type tasks may be limited to those periods when the condition is in its clinical phase. During the clinical phase, both schizophrenics and depressives may experience extreme difficulty in making correct judgements about the intentions and beliefs of the people with whom they interact, even though they have quite normal mind-reading abilities when they are in remission from the condition. This may explain why paranoia is often an important component of schizophrenia: they completely misconstrue the intentions of the people with whom they come into contact and thus believe, wrongly, that these individuals are scheming against them. It is as though their mind-reading modules are working overtime without the benefit of the more rational part of their mind to act as a brake on the more outrageous inferences made on the basis of tenuous evidence.

There is one other group of humans who also score poorly on theory of mind tasks. These are the individuals we commonly refer to as autistic. Autism is a strange and unnerving condition that, although fortunately rare in the population as a whole, is three times more common among boys than among girls. Its main defining feature is the lack of theory of mind, though there may be a number of associated symptoms across the wide range of conditions defined as autistic (including variations in IQ, language ability and many other cognitive skills). At one end of the spectrum (usually referred to as Asperger syndrome), individuals may be of normal or even above normal IQ; at the other, autistic individuals may suffer severe and chronic learning difficulties and may even lack language. What unites this diverse range of conditions more than anything else, however, is the fact that all these individuals fail false-belief tasks – they lack theory of mind.

In this context, Asperger individuals are particularly interesting, since they are typically of normal or even above-average intelligence. Many are often especially good at mathematics and computing. The cost they bear is that they cannot deal effectively with the social world in which we are all, perforce, obliged to live. They do not understand how or why people tick, and invariably offend or alienate friends and acquaintances by their uninhibited and direct ways of interacting. In other words, they do not understand the subtleties of normal social interaction – that intuitive appreciation we have of knowing just how far to push things, just what are the right nuances of wording that will convey what we want to say without actually saying it, or how to make a subtle hint that doesn't *quite* commit us to anything if it is rejected, or a polite brush-off that allows the recipient to retire gracefully without feeling that they have been publicly humiliated.

This failure to pass ToM tasks even in adulthood has a number of implications for autistic individuals' social lives. Because they lack ToM, autistic children do not lie (or at least cannot lie convincingly) and they do not engage in pretend play. They do not understand what it means to pretend that a doll is alive and might be hungry or sad. They take the world exactly as it comes: animate things are animate and inanimate things are inanimate, and that's that. By the same token, when they develop language, they use words only with their literal meanings. They do not understand what it means to joke or to use words metaphorically. As a result, it is necessary to be especially careful when speaking to autistic individuals: they will take you exactly at your word. An instruction like 'Pull the door behind you when you go out' will be taken to mean *exactly* what it says – remove it from its hinges and drag it behind you.* That interpretation simply does not cross our minds because we are so used to the word 'pull' being used in a metaphorical sense in these contexts. We know instinctively from the context whether the speaker really intends us to yank the door off its hinges and drag it behind us – or just to close it. There are occasions when speakers do mean the first of these, and occasions when they mean the second, and we learn to interpret the speaker's intentions (his or her *mind state*) on the basis of appropriate contextual cues – and, no doubt, a certain amount of inspired guesswork.

Lack of theory of mind has devastating implications for an autistic individual's ability to manage social relationships. Such people simply do not understand how to manage the subtleties of everyday social discourse. Even as adults, they persist in mak-

* This real example was described to psychologist Francesca Happé by the mother of an autistic son to whom it happened.

ing mistakes and this makes it difficult for them to maintain relationships through those subtle give-and-take arrangements that the rest of us seem to execute so effortlessly.

If half of you have the slightest suspicion that you might have encountered this kind of behaviour more often than you might wish in real life, you may not be so far wrong. The psychologist Simon Baron-Cohen has argued that, in reality, autism is part of the normal syndrome of maleness in our species, carried to extreme form in a few unfortunate individuals but simmering quietly beneath the surface of every male mind. There is a lot to be said for this view. Not only is it a commonplace of folk wisdom that women are more sensitive to social cues and much better at handling social situations than men, but there is a great deal of evidence to support the claim. A study by one of my students, Rebecca Swarbrick, showed that women are indeed statistically better at solving second- and third-order theory-of-mind tasks than are men. There is, of course, a great deal of overlap in the two sexes' abilities – some men are better than some women. But, on average, women are much better than men at these kinds of tasks. We see the effects of this in everyday life. Think of how intense the social relationships of pre-teenage girls are. For a girl, relationships are often so intense and focused that not being invited to Penelope's party is little short of the end of the world; for boys of similar age, a relationship largely involves kicking a soccer ball back and forth across a street – a wall would do almost as well as another boy, providing it returned the ball.

'Knowing How' and 'Knowing That'

This appreciation of the peculiarly complex nature of human mental life leads us to an inevitable question: is theory of mind unique to humans or is it a more general feature of animal psychology? Are we alone in the extraordinary mental universe in which we live?

It has not proved easy to answer this deceptively simple question. One of the difficulties is that we are so immersed in our own quite remarkable mental world that we naturally ascribe these same abilities to the rest of the animal world. Indeed, so natural is it for us to interpret other organisms' behaviour in this way that we even ascribe mental states to the inanimate world. We believe that trees and springs are (or have) spirits. Our very language is redolent with mentalised descriptions of the physical world: we speak of 'threatening clouds', 'angry seas', 'the wind sneaking in through the gaps in the door', 'raging storms'. This tendency to anthropomorphise is so much a part of our psychology that we have to guard against it very carefully when we are trying to understand the world in which we live.

Believing that other animals or even inanimate features of the world have mental states like ours is a convenient and harmless phenomenon in everyday life, but is apt to mislead us badly when it comes to understanding the way the world in which we live really works. If we insist on ascribing human motives to volcanoes, for instance, we are misled into trying to cope with their occasional ravages by making placatory sacrifices rather than by trying to understand and control the physical forces that actually drive them. Rather than trying to learn how to predict their periodic outpourings, we instead resort to useless prayers and grovelling entreaties. In the world of the hunter-gatherer, that might be a harmless activity, given that there is precious little that any human can do to prevent natural disasters of that kind. And it might even help us get through the occasional calamity by making us feel more in control and so more able to cope with the uncertainties of the world. But in today's world in which so much of what we do is interdependent and itself dependent on technology, prayer may only make things worse by distracting us from what we really can and should be doing. Computers are, after all, zero-order intentional beings and, try as hard as they might, they are notoriously unable to understand our intentions.

The issue of anthropomorphism, then, is a very real one. If we want to be absolutely sure that animals have mental abilities similar to our own, then we have to put them through tests that do not fall foul of what psychologists have come to know as the Clever Hans effect. Kluge ('Clever') Hans was a horse from turn-of-the-century Germany whose stock-in-trade was the ability to count. His owner, the sometime Russo-German aristocrat and retired teacher Wilhelm von Osten, enthusiastically toured Germany with him. At these displays, von Osten would call out to the horse: 'Hans . . . what do three and four add up to?' Hans would start tapping with his forefoot while von Osten counted out the taps. One . . . two . . . three . . . four . . . and after the seventh tap, Hans would stop. The audience would be amazed. No matter what numbers the horse was given, he invariably got the answer right. For von Osten, Hans provided evidence to support his own pet theories about education and he touted the horse around the country in the hopes of promoting these.

Like his audiences, von Osten was completely convinced that his horse could count, so he was willing to put his protégé to the test at the hands of scientists. Eventually, after a long series of careful experiments, it turned out that Hans was not counting at all, but picking up very subtle cues from Herr von Osten. By carefully manipulating what the horse could see and who was in the room with him, the experimenters were able to show that, as the horse hit the last tap in the sequence, von Osten gave a distinctive nod of the head, as though to say 'That's it! There you are!' Hans picked that cue up and stopped tapping. Von Osten did not do this intentionally. It was just the natural reaction for anyone who counts slowly with a child. But the horse had learned that it got rewarded whenever it stopped tapping on that cue. What we see is not always what actually is. The spectre of Clever Hans continues to haunt psychologists to this day.

A number of attempts to devise foolproof false-belief tasks for use with animals have been made in the last few years, though so far they have only really been tested on chimpanzees and dolphins. Some of these tests have been designed to determine whether the animals can discriminate states of knowledge from states of ignorance in another individual: in other words, whether they understand that *seeing* implies something about *knowing*. Danny Povinelli, a psychologist at the University of Louisiana in the USA, offered chimpanzees and rhesus monkeys the opportunity to choose between carers who were knowledgeable and those who were ignorant or who behaved in an unkind way, whether deliberately or accidentally (e.g. spilling the drink that the animal had been offered). Chimps learned quickly to request a food reward from a carer who was facing them rather than one whose back was turned, but they were less reliable at distinguishing between one facing them and one whose head was covered by a paper bag. In another experiment, a carer had to tell the chimpanzee in which of two possible places a food reward was hidden. The chimpanzees learned quite rapidly to respond to instructions given by a carer who had watched the food item being hidden and to ignore another carer who had been out of the room at the time (and hence had to guess where the food was). In contrast, rhesus monkeys did poorly on these tasks, suggesting that apes, but perhaps not monkeys, are able to interpret states of knowledge and ignorance correctly.

However, Povinelli remains sceptical as to whether even the apes were doing more than learning very fast which was the more reliable cue as opposed to understanding the carer's mind-state. One observation that makes him particularly sceptical is that the chimpanzees responded at no better than chance levels on some of these tests when re-tested two years later. It is as though, having once rote-learned the rule for solving the problem, they had now forgotten it. The implication is that they do not have a natural ability to understand others' mind-states, but rather, like Asperger people, they learn how to solve the problem using more superficial cues. Since each problem is unique, the solutions are quickly forgotten. In contrast, theory of mind provides a more general high-level rule that allows different kinds of social situations to be solved using the same basic principles.

However, knowledge and 'seer-knower' problems are not quite the same thing as false-belief tasks, which is why developmental psychologists have used these as the benchmark for ToM in children. False-belief tasks can *only* be solved correctly using theory of mind. How might animals fare on these kinds of tests?

Josep Call and Mike Tomasello presented chimpanzees with a simple false-belief task. In this task, the chimpanzee was asked to choose between a pair of boxes, one of which had been baited with a piece of fruit behind a screen. The only clue the chimpanzee had available to it was the fact that a human observer had watched the baiting behind the screen and later indicated which box to choose by tapping on the baited box after the screen had been removed. The chimpanzee learned to rely on the observer as an honest broker. Once the chimpanzee was performing reliably on this task, a new element was added. After the boxes had been baited, the observer left the room; while she was outside, the experimenter swapped the boxes over. To be able to solve the problem now, the chimpanzee needed to understand that, when the observer returned and tapped the box, she now had a false belief because she did not see the boxes being switched. The animal should now choose the *opposite* box to the one the observer tapped. But if chimpanzees lack ToM and simply learn a behavioural rule ('Always choose the box tapped by the human'), it will continue to choose the box indicated by the now 'confused' observer. In this experiment, young children passed this test with no trouble, but the chimpanzees comprehensively failed.

In another study, Sanjida O'Connell (then one of my PhD students, now a TV producer and award-winning novelist) tested chimpanzees on a similar four-box problem that had been benchmarked against the Smartie test using four- to six-year-old children. Although the chimpanzees did significantly better than chance on this task, they clearly did not perform anything close to the level of five- and six-year-old children. They did about as well as four-year-olds – in other words, children who are just on the brink of acquiring theory of mind. However, perhaps the more significant finding was that the chimpanzees did significantly better than autistic adults on this task.

In effect, we have one strike in favour (but only just) and one strike against the possibility that chimpanzees understand false-belief tasks. The fairest conclusion is perhaps that the jury is still out on whether or not chimpanzees have theory of mind. But it seems that, if they do, then, at best, they can only just aspire to second-order intentionality – the level of performance of a four-year-old child, and three full levels down on what normal adult humans can manage.

Dolphins have exceptionally large brains, and their convoluted shape with lots of wrinkles in the surface makes them look more like humans' brains than apes' brains do. They thus seem an obvious candidate for species that might have mind-reading capacities. Indeed, we know that their social behaviour is very complex. They form coalitions much as monkeys and apes do, and they have long been credited with the ability to understand (and, some have claimed, use) language. We therefore tested dolphins with a version of the same two-choice task used by Call and Tomasello. In the first series of experiments (carried out in Durban, South Africa, by Al Tschudin), the results were very encouraging. Dolphins seemed to be able to pass false-belief tasks with astonishing ease. But we became worried about the possibility of a Clever Hans type effect: careful analysis of the videotapes of some of the experiments suggested that the dolphins might have been using inadvertent cues provided by the experimenters to solve the task. It was also clear that the dolphins frequently put pressure on the experimenters by hedging their bets and trying to indicate both boxes at once: they would point their noses at one box as they were required to do, and then drift gently across towards the other box, so beguiling the experimenters into giving them the benefit of the doubt as to which box they had actually meant to point to. Dolphins are, of course, exceedingly smart animals; more important, perhaps, in this particular case, the animals that took part in the experiment were used to working with human trainers in public shows. When the experiments were repeated with a naïve set of dolphins (this time in Florida), a new set of experimenters (Elainie Madsen and Heidi Feld) and a more rigorously designed experimental procedure that carefully ruled out both Clever Hans cuing effects and bet-hedging by the dolphins, the dolphins quite comprehensively failed to pass.

One reason for these inconclusive results on non-human animals could be that the tasks are too artificial for either the chimpanzees or the dolphins to master. Perhaps it is not that they cannot do theory-of-mind tasks, but that they cannot figure out what the point of this particular game is, irrespective of whether it requires theory of mind to solve correctly. All these false-belief tasks suffer from two features that might make them difficult for apes and dolphins to get their minds around. One is that the tasks essentially involve cooperation between the subject and the experimenter (or demonstrator). Animals who are not naturally especially cooperative might find it difficult to engage in a cooperative task with another individual when the task involves something like food that they normally compete for. A task that was more competitive might be more natural for an animal, and therefore easier for it to understand. The second potential problem, of course, is that all these tasks involve interactions with humans. Apes and dolphins might find trying to understand a human mind much more difficult than trying to understand another ape's or dolphin's mind.

The Harvard psychologist Brian Hare took these lessons on board and devised a more chimpo-centric task. This task involved placing a food reward (say, a piece of fruit) in the space between two cages occupied, respectively, by a dominant animal and a lower ranking one. When the two animals were released into the space, the subordinate chimpanzee typically hung back and allowed the dominant animal to get the reward. However, when the piece of fruit was placed behind a solid screen so that only one chimp could see it, rather different results were obtained. If it was placed so that the subordinate could see the fruit and the dominant could not, the subordinate quickly went to the food and ate it. It seemed to understand that it could see something the dominant animal couldn't, and that it could steal a march on the dominant before the latter realised what was going on. This is not *quite* the same thing as passing a false-belief task, but the ability to take another's perspective is a major step in that direction. Once again, these results seem to suggest that apes (or at least chimpanzees, since no one has yet tested gorillas or orang-utans) might be hovering on that crucial theory-of-mind boundary.

Even allowing for the most charitable interpretation of these results, however, one thing seems clear: neither chimpanzees nor dolphins perform anything like as well as six-year-old children, who most definitely do have theory of mind. And whatever it is that apes can do, they are simply not on the same scale as adult humans in terms of the latter's ability to cope with fourth- or fifth-order intentional problems.

On one thing everyone seems to agree, however, and this is that, in contrast to the apes, monkeys genuinely do not have theory of mind. Although conclusive tests have yet to be carried out, there is considerable circumstantial evidence to suggest that they would fail them. Dorothy Cheney and Robert Seyfarth spent many years studying vervet monkeys in Kenya's Amboseli National Park. One day, they watched the dominant male of their study group showing concern about a strange male hovering in a neighbouring grove of trees. It was clear from his behaviour that this strange male was intent on joining the male's group; if he succeeded in doing so, there was more than a fair chance that the group male would be demoted from his exalted position as the dominant male – and thus denied the privilege of mating with the females when these came into breeding condition. The male was presumably pretty unhappy about this prospect, but hit on the ingenious plan of giving 'Beware – leopard!' alarm calls whenever the strange male descended from his grove of trees to try to cross the open ground that separated him from the group. Vervets, in common with many species, have different alarm calls for different kinds of predators. These calls identify the source of the danger, and the other animals respond in appropriately evasive ways whenever they hear these calls – dashing for the nearest tree when they hear a leopard call, diving down out of the tree canopy when they hear an eagle call and standing up to peer into the long grass when they hear a snake call. Whenever the group's male gave a leopard call, the strange male did just what he was supposed to and fled back into the safety of his own grove of trees.

All was going well until the male made the mistake of giving his leopard call while himself nonchalantly walking across the open ground between the two groves of trees. The game was immediately up. It was obvious even to the most dull-witted monkey that there was no leopard. Had there really been one present, the male would not have been wandering so laconically across the open ground. It's a bit like the three-year-old child who insists that it hasn't eaten the chocolates from the fridge, knowing that past experience suggests that if you say such things with sufficient conviction there is a fair chance that adults will believe you. But a three-year-old lacks theory of mind, so it doesn't realise that adults are convinced only because they are willing to give the little scamp the benefit of the doubt . . . and benefit of the doubt is not an issue when its face and hands are covered in chocolate.

In Cheney and Seyfarth's terms, the difference between monkeys and apes might be that between being a good ethologist and a good psychologist. Ethologists study behaviour and are good at interpreting the meaning and significance of behaviour, at least in terms of predicting what is likely to happen next. Monkeys are able to use this kind of knowledge both to predict what others will do and to manipulate them, but they do not understand the mind that lies behind the behaviour: they do not mind-read, and therefore they make crucial mistakes because they do not understand *why* the other animal behaves the way it does. Psychologists, in contrast, interest themselves in the mind behind the behaviour and can thus exploit more sophisticated opportunities for deception. It is an old distinction that philosophers of mind draw between *knowing that* and *knowing how*. I can know *that* something happens, but I may not know *how* that effect is produced.

The Deceiver's Art

Monkeys are clearly experts at manipulating their fellows, just as humans and apes are. Dick Byrne and Andy Whiten of the University of St Andrews collated a large database of examples of deception from the scientific literature on primates. All these cases involved one animal attempting to manipulate the behaviour of another. Because they were putatively doing so by manipulating the stooge's knowledge of the situation, Byrne and Whiten referred to these kinds of phenomena as 'tactical deception'.

An example of this kind of behaviour was witnessed in a group of hamadryas baboons, which inhabit the dry deserts of the north-eastern corner of Ethiopia. This species lives in large troops of eighty or so animals that range widely over the thornscrub along the desert's edge. Within each troop can be found a number of smaller units, each containing a single breeding male and two or three females, plus their dependent young. The males guard the integrity of their little units with considerable ferocity, punishing their females if they become separated from the unit or show the slightest indication of interacting with other males in the group. The Swiss zoologist Hans Kummer, who has done most of the work on this species in the wild, once watched a young female spend twenty minutes inching her way towards a large rock while her unit was busy feeding. Behind the rock lay a young adult male who had no females of his own. Once behind the rock, the female began to groom the young male. But, according to Kummer, she seemed to make every effort to ensure that her head was above the rock and clearly visible to her own male feeding a dozen metres away. It was as though she was thinking: 'So long as the old codger can see my head above the rock, he will think I am innocently sitting here . . .'

Intriguing as these observations are, they suffer from the worry that most of these examples are open to alternative interpretations. Consider the example of the hamadryas female. On face value, the female seems to have in mind the idea that so long as her own male can see her head above the rock, he will suppose that she is behaving herself: she *thinks* that the male will *believe* that she is innocently feeding. It looks remarkably like second-order intentionality, enough to justify our attributing theory of mind to monkeys. But we cannot be sure that this is what she really is thinking. Maybe she is simply being a good ethologist, having learned the hard way that if she does not keep in direct view of her male, he will punish her. She does not really understand why the male behaves in this way, but she knows enough to appreciate that the best way to avoid being attacked is to stay well in view. Or perhaps she is simply thinking that if she keeps an eye on the male, she will have sufficient warning of an attack to be able to escape to safety.

We are left with two equally plausible explanations for what is going on, both of which would explain why the female behaves in the way she does. The charitable explanation assumes that she has theory of mind and can deliberately deceive her male by feeding him false information; the less charitable one assumes that she has simply learned how to avoid trouble without really understanding how and why the strategy works as well as it does. On the basis of the female's behaviour alone, we cannot tell which explanation is the right one. But which option is right makes a huge difference to our understanding of monkeys' minds.

One could argue that attributing mental states is the most efficient way for any animal to learn to avoid punishment. If an animal simply learns that it gets punished in certain situations (i.e. it learns that specific consequences often follow after certain events or cues), it will find it hard to avoid incurring punishment by accidentally being in the wrong place at the wrong time on some future occasion. Its life will always consist of a series of punishments that it will find difficult to predict and control. On the other hand, if it can apply theory of mind to the problem, it will be able to work out what kinds of situations incur the wrath of the other animal(s) because it will understand just what it is that bothers them about its behaviour (presumably, in the hamadryas female's case, the fact that her male is worried he might lose her to another male). It will then be more effective at anticipating future problems and so be able to do something to pre-empt them.

This is (for us, at least) a natural way of dealing with the social world. It is also one that works very efficiently. We clearly have the computing power in our brains to do all the necessary calculations. But this does not mean to say that all other animals have the same abilities. It seems obvious that at least some animals do get by quite adequately in the world without resorting to these kinds of mental gymnastics. Insects are one obvious example, fishes and reptiles are probably another; mice and rats might be a group a little closer to home. Of course, all these species live in much simpler social worlds than monkeys and apes do. So the issue really boils down to the question of whether monkeys and apes *need* theory of mind to work effectively in the kinds of societies they live in. Is the kind of social complexity seen in monkey or ape societies *only* possible if the animals have theory of mind – or are human relationships *that* much more complex than theirs?

But perhaps we are missing something here? Our focus on theory of mind may be obscuring something much more important about the way we humans work. Perhaps theory of mind itself is simply the emergent property of some much more fundamental properties of how the human mind ticks.

Deep Thought

One of the very conspicuous features of our mental world is the way we rehearse what we are going to do. This often entails explicitly considering alternative options, evaluating their likely outcomes and, having chosen one, rehearsing how we can best approach its execution. This process is so much a part of our mental life that we barely give it a moment's thought. But perhaps it provides us with the clue we have been looking for.

This kind of mental rehearsal is actually quite a complex task and involves bringing into play a number of quite different cognitive abilities. At the very least, these include the ability to reason causally (to follow the sequence through from cause to likely effect), to reason analogically (to recognise that A is to B as X is to Y), to run several alternative scenarios in parallel and, finally, to do so on an extended time frame into the future.

Analogical reasoning may be the dark horse here. Lera Boroditsky has in fact recently suggested that we use it to develop a sense of time off the back of a sense of space. Time, after all, is something that we can only imagine: we cannot touch or sense it directly. Instead, we reflect on memories of past circumstances, compare them with the current state of the world, and infer that time has passed. In contrast, we have direct sensory experience of the world of physical space through both vision and touch. Boroditsky showed that, when primed with statements about spatially arranged objects ('The flower is in front of me'), subjects were more likely to interpret an ambiguous statement about the temporal arrangement of an event ('Next Wednesday's meeting has been moved forward two days: which day is it now on?') in ways consistent with the spatial arrangement that they had been shown; however, a statement about a temporal event did not have the same effect on how they interpreted an ambiguous statement about a spatial arrangement. She argues that we develop a sense of time analogically from our sense of space, and this explains why we use so many spatial metaphors when referring to time. Things happen *before* other things; we look *ahead* or *forward* to the future; we fall *behind* schedule.

Analogical reasoning may play an unexpectedly crucial role in the story of the human mind because it provides us with a platform for understanding other minds. I use my experience of my own mental processes to imagine how someone else's mind might work. I use it to empathise with your predicament when the ATM swallows your cash card, as well as with your feelings when you stub your toe. This phenomenon may be even more important when it comes to dealing with the world of relationships. We can experience directly the interactions that pairs of animals engage in: we see these happening before our eyes, we may even experience them directly through touch or hearing. But when it comes to understanding the nature of the *relationship* between two individuals, we have to deal with something that we cannot experience directly. As the ethologist Robert Hinde pointed out more than three decades ago, we abstract relationships out of the observations we make of individuals interacting. Relationships are things that happen only in a virtual world, and we have to be able to move backwards and forwards between the physical world of interactions (real events) and the virtual world in which these events are constituted into relationships in order to be able to understand what the significance of specific actions is or might be, or how two relationships impinge on each other.

Testing animals' abilities to engage in analogical reasoning is not exactly easy. Most attempts to explore this phenomenon have focused on the rather simpler task of whether they understand analogies between categories ('Tap is to water as key is to . . . [lock]?'). This is not quite the same thing as being able to exploit the analogy between how, say, one mind works and use it to model the behaviour of another mind. Or to take a social process and use it to model a physical process, or *vice versa*. All the studies of this phenomenon so far have concentrated on *perceptual* similarity rather than *conceptual* similarity. Conceptual similarity may be more important in dealing with the social world because relationships cannot be experienced directly. Attempts have been made, however, to test great apes' abilities on at least two of the other processes, namely causal reasoning and mental rehearsal.

The test we used to assess an understanding of causality is one that has been used extensively on very young children – even babies as young as six months. The design is relatively simple: a subject (human or otherwise) is shown a video clip of, say, one object hitting another, thereby causing it to move. The sequence is repeated over and over again until the child habituates (ceases to pay attention); it is then shown a clip in which movement occurs despite the fact that the two objects did not touch (or 'collide'). If it suddenly pays attention, this is taken as evidence that it appreciates that something odd or out of the ordinary has happened (at least, providing it can be shown that it does not respond as strongly to a simple change of scene). This is taken as evidence of an intuitive understanding of causality. Children as young as six months pass this test, and so it seems do chimpanzees – but not monkeys. Rather similar results using different kinds of causal reasoning tests have been obtained by the Italian psychologist Elisabeta Visalberghi and her collaborators in Rome.

We tested apes' ability to engage in mental rehearsal by comparing the time taken to open a puzzle box after the animals had had an opportunity to look at (but not touch) various puzzle boxes for a day with that when they were given the puzzle box without any rehearsal time. Chimpanzees, orang-utans and young children (aged five–seven years) all did much better after an opportunity to think about a box than they did when given the box cold. But, significantly, even such young children were a great deal faster on the task than either of the great apes.

These results suggest that such basic abilities may be quite widespread, at least among the apes. Nonetheless, it is clear that however good the apes are on these kinds of tasks, human children are simply orders-of-magnitude better, even at very young ages. Two things are likely to be important here. One is that, for full-blown human-style social cognition, all four abilities have to be brought into play together: having only some of them is useful, but it does not allow you to engage in the kinds of complex thinking that goes into fourth- and fifth-order intentionality. Second, the extent to which these abilities can be deployed probably depends directly on the size of your computer (i.e., brain): how many video tapes you can run side by side, and how far into the future you can run them, may be a simple function of how much neural circuitry you can afford to devote to the task (and still have enough left over to keep the rest of the body going). Planning depth may prove to be especially important in this context: it is what allows us to play chess, as well as to scheme our way into other people's good books. My colleague Louise Barrett, who studies baboons in the wild, always insists that the trouble with baboons is that they live wholly in the here-and-now and do not seem to be able to figure out that sometimes holding back on the emotions of the moment might allow them to do better in the long run.

The ability to step back from the immediacy of the world may be crucial in allowing us to assess the consequences of alternative courses of action. This, after all, is in effect what theory of mind is all about – the ability to step back from one's personal experience and imagine that the world could be other than it is, to imagine that someone else could have a false belief about the world. Differences between species in these very basic abilities may explain why baboons and great apes are very competent socially but autistic children are not, even though none of them can pass a false-belief task.

Brain Story

Phineas Gage has the rare distinction of having achieved immortality. Sadly, it is probably not the kind of immortality he might have had in mind had he given a moment's thought to it. So far from being physically still with us or remembered for the fine symphonies he composed or the exquisite paintings he produced, he lives on as one of the most celebrated cases in neuropsychology, familiar to generations of psychology students who continue to learn the bare bones of his life story a century and a half after he himself died.

Phineas had been the foreman in charge of a road gang laying tracks for the new railway near Cavendish, Vermont, in north-eastern USA. He ran a good crew, among whom he maintained discipline and one of the best work rates on the line by dint of a forceful personality and an ability to cajole and persuade – no mean feat, given that road gangs were mostly made up from a hardbitten and fractious bunch of social misfits. Then, one fateful day in September 1848 while he was preparing a charge of explosives to blast through a rocky cutting, the gunpowder accidentally ignited as he was tamping it down in the hole in the rock with a three-foot-long metal rod. The force of the explosion drove the tamping iron straight up through the front of his skull, destroying a large chunk of frontal cortex.*

* Since his skull was preserved after his death, it has been possible to use modern computer methods to model the passage of the tamping iron through his brain, allowing us to determine exactly which bits of his frontal cortex were destroyed and which spared.

Miraculously, Gage survived. But, as he recovered, his personality changed completely. The once masterful manager of a crew of tough, hard-drinking labourers was no longer able to maintain effective social relationships and became chronically incapable of getting a job done. He was unpredictable and inconsiderate, constantly blaspheming and behaving in an impatient, obstinate and capricious manner. According to some (albeit possibly apocryphal) accounts, he died penniless in an institution, having drunk himself to death twelve years after his accident.

Gage's importance in the history of neuropsychology lies in what his unfortunate accident has to tell us about the functions of the frontal cortex, that bit of the outer layer of the brain that lies above the eyes and, roughly speaking, forward of the ears. We have long appreciated that herein lies the seat of conscious mental activity, the part of the brain that is intimately bound up in all those smart activities that we especially associate with humans. Yet, Phineas Gage's experience tells us that we can survive pretty well without substantial chunks of this bit of the brain. We just do not need it for day-to-day survival. Gage, after all, lived for a dozen years after his accident and, while the later part of his life may have been less fulfilling than his early history might have promised, it was by no means unhappy. He, at least, seems to have been pretty contented with his life in his later years, even if others were not quite so enthusiastic about the way he treated them. What it does suggest, however, is that there is something special about this chunk of the brain that plays a crucial role in smoothing our way through the bumpy vagaries of our social world.

Phineas Gage's sad story reminds us that much of what we do in the social domain is a fine-tuned balancing act teetering precariously on the brink of social disaster. That most of us humans manage to keep to the side of social cohesion is largely thanks to the frontal cortex of our brains. Whatever the psychological processes involved may be, they must in the end be consequences of brain activity. Consciousness, as we experience it, is nothing more than the emergent property of electrical activity in the brain as interconnected neurones exchange electro-chemical messages. We can reflect on these events (the phenomenon we call 'self-consciousness') because we have theory of mind and can stand back from our immediate experiences and ask how it feels to think something. In other words, we can ask, how do I *know* that I *know* something is the case? But how is it that only we humans can do this?

A hint as to where the answer to this question might lie comes from the observation that social group size in primates correlates with the relative size of the neocortex of the species concerned. The neocortex is the relatively thin sheet (about 6 millimetres deep) that is wrapped around the inner core of the ancestral reptilian brain that all vertebrates share. In mammals, this sheet typically accounts for between 10 per cent and 40 per cent of total brain volume, but in primates it starts at 50 per cent (in the prosimians) and rises to as much as 80 per cent of total brain volume in humans. Put somewhat simplistically, the neocortex is the thinking part of the brain and big neocortices are *the* primate speciality.

It is important to notice that, during the course of primate evolution, the brain has expanded forwards from back to front, so that the bit that has increased out of all proportion in modern humans is the frontal lobe. The bits at the back and sides of the brain are mainly devoted to vision and other aspects of sensory perception, sensory integration and memory. It is the increased size of the frontal lobes that is largely responsible for the much greater intelligence of species like apes and humans. Of course, this is not the whole story, since in reality the brain is a highly integrated organ, with complex interconnections between different parts of the neocortex as well as between the neocortex and some of the more primitive parts of the brain (notably some parts of the limbic system, which deals with emotions and responses to emotional cues). However, this simplified picture provides us with a good enough basis for understanding some of the key cognitive differences between humans and other primates.

There is a correlation between social group size and the volume of the neocortex in primates which suggests that it has been the need to manage the complex social world in which primates live that has driven the evolution of ever-larger brains.* The important point for the present story is that we humans fit neatly onto the same scale as the other primates. Group size in humans is about 150: this is the number of people that you know personally and have some kind of meaningful relationship with – as opposed to the people you know by sight or those with whom you have a strictly business relationship. Chimpanzees live in communities that have an average size of about 50–55, and their neocortex is proportionately smaller.

However, it turns out that, as brain size has increased during the course of primate evolution, the various parts of the neocortex have not expanded in the same proportions. The sensory processing areas of the neocortex seem to increase in size less fast than the non-sensory components in the frontal lobe. This is mainly because there is no advantage in having a computer to analyse sensory input (the information that comes in through your eyes, ears, nose and so on) that is bigger than that minimally necessary to make sense of the signals from the relevant organs. Given that the neocortex as a whole is increasing at a much faster rate, more spare capacity becomes available for the smart stuff that goes on up at the front end – in other words,

* I discussed this, and its implications for the evolution of language, in considerable detail in my book *Grooming, Gossip and the Evolution of Language*.

social skills such as theory of mind – as brain size increases across the range of sizes found in the monkeys, apes and humans. The amount of spare capacity in the frontal lobe over that available to monkeys first begins to show a significant increase at around the brain size of great apes (which may explain why they, but not monkeys, can just about manage theory-of-mind tasks), but it is more than four times greater in modern humans than in great apes and the rate of increase seems to be exponential.

There is now quite a lot of clinical evidence to support the suggestion that the frontal lobe of the brain might play the crucial role in mind-reading. Patients who have suffered lesions in the frontal cortex as a result of accidents or strokes, for example, invariably lose their social skills. In some cases, they merely lack the ability to execute the usual social graces and behave like people with Asperger's syndrome, trampling unwittingly on others' social sensibilities without embarrassment; in other cases, like the unfortunate Phineas Gage, their entire personality may change and they become more aggressive and less considerate of others' interests.

Recently, new technology has allowed us to peep inside the brain while it is actually working. The technology depends on the reasonable assumption that when bits of the brain are actively working on a problem, they consume more oxygen than the resting brain and so blood-flow to those particular points increases. Blood-flow through small segments of brain can be measured indirectly using the changes such activity creates in the electromagnetic fields that surround the brain, or in the frequency with which electrons are emitted, both of which can be picked up by powerful recording devices. Studies of the active brain suggest that various areas up in the frontal cortex are particularly active when we are engaged in thinking about social cognition tasks like the Sally-Ann task, but not when we are thinking of simpler tasks like recognising shapes or reading words.

Taken together, these results suggest that, as the ape and human brain has evolved in size, the extra capacity has largely been added on at the front where it can be put to use developing more powerful social cognitive abilities. Eventually, at some point in hominid evolution, sufficient extra computing power was available to make that crucial transition into the kind of cognitive reflexivity that allowed us to engage in second- and third-order intentional analyses of the world we lived in.

Inevitably. the obvious question at this point is: When did our ancestors pass through the critical Rubicon at which theory of mind and higher orders of intentionality became possible? The short answer, of course, is that it is rather hard to say because neither brain nor behaviour, let alone mental states, fossilise terribly well. However, we can gain some idea by relating the kinds of findings I've discussed above to the changes in brain size in the hominid lineage, shown in Figure 2 (p. 29). We can do this because scaling relationships within the brain mean that overall brain volume gives us a reasonable idea of the relative sizes of the constituent bits.

If we map the intentionality levels of monkeys (at first-order), apes (at second-order, just) and modern humans (at fifth-order) onto the relative size of their frontal lobes, we get a surprisingly good straight line relationship. Using this relationship, we can discover what neocortex size would equate with third-order intentionality, and then find out when the equivalent size of brain appeared in the human fossil record. Since one of the things that can be determined reasonably well from fossil specimens is brain volume (the skull, being especially hard, tends to be preserved better than most bones), it should be possible to map onto the history of hominid evolution the pattern of change in these crucial mentalising abilities.

Mapping this relationship onto the graph of changes in brain size in the hominid lineage shown in Figure 2, making the necessary adjustment for a very straightforward relationship between total brain volume and the volume of the frontal lobe, gives us the results shown in Figure 6 (p. 191). These suggest that third-order intentionality would have appeared for the first time with *Homo erectus*, around two million years ago. Fourth-order intentionality, however, would not have made its appearance until sometime around 500,000 years ago when archaic *Homo sapiens* (our own species) came on the scene. Because brain size continues to increase dramatically in the human lineage, fifth-order would have followed fairly quickly on its heels. It is worth noting that both the Neanderthals and the Cro-Magnons, like contemporary humans, had brain sizes large enough to accommodate fifth-order intentionality. It seems that the Neanderthals might not have been the intellectual slouches of common myth.

It seems that although the critical first step into higher levels of mentalising was made quite early on, the critical ones that radically distinguish us from our ape cousins – the higher orders of intentionality – probably came in very late: at the earliest, with the appearance of *Homo sapiens*. Whether or not the Neanderthals shared these capacities with us really depends on whether their brains were organised in exactly the same way as ours. The famous Neanderthal 'bun' (the enlarged back part of their skull) suggests that they might have had a much bigger visual area than we do (something that is confirmed by the relatively much larger size of their eyes); if so, then it is possible that they would have had less neocortex volume in their frontal lobe – which, if true, could have limited their social cognitive abilities to fourth-order intentionality (the level they would have inherited from the archaic humans that we and the Neanderthals share as a common ancestor). If this is so, then fifth-order intentionality, and all the complex social phenomena that depend on this, would not have appeared until anatomically modern humans (our own sub-species) came on the scene a mere 200,000 years ago.

The conclusion we have been drawn to in this chapter is that, although apes and humans share a number of important advanced cognitive abilities, they differ in one key respect: the extent to which humans can detach themselves from the world as they experience it. This allows humans to reflect on the world as they find it, to wonder whether it could have been otherwise. In contrast, apes (and certainly all other animals) have a much more direct, straightforward experience of the world. Their noses are thrust firmly up against reality. In the following chapters, we shall see that this has very important implications for some of the more explicitly human aspects of our behaviour.

4 Brother Ape

Godi tensed when he heard the faintest crack of a dry twig behind him. He looked carefully round, peering uncertainly into the bushes. The silence was oppressive. In the distance, a bell bird began to call its sombre solitary note at intervals, awaiting its mate's reply. Nothing else stirred. Godi relaxed slowly. Perhaps he was mistaken. He turned back to his interrupted eating. But as he did so, a shadow moved across the corner of his eye.

Suddenly, from all around him, forms condensed out of the concealing foliage. Godi knew at once what these meant. He had been ambushed by the Kasekela males. He leaped for the tree branches above him, desperate to out-climb them. But it was too late. In his panic, Godi missed his grip and fell. Momentarily winded by the fall, he was not quick enough to get up and lay shocked and rigid as the screaming males, each with the strength of several human men, encircled him and began to pummel and stamp on him in a chaotic yet concerted attack.

The attack continued for twenty minutes before the screaming posturing males, the hair on their shoulders bristling, withdrew back into the shelter of the surrounding trees. Then, as silently as they had come, they disappeared northwards back into their own part of the forest.

Godi lay stunned and shocked, his body broken and bleeding,

his head humming and confused. After half an hour, he dragged himself into a sitting position. His chest and head hurt as they had never hurt before, and one arm hung uselessly at his side, broken when one of the males had pulled viciously on it with all his strength. Slowly, and with painful care, Godi began to drag himself towards the stream in order to slake the raging thirst that gripped him. He died of his internal injuries a day later beside the stream.

And so began one of the most shocking discoveries of the 1980s. The males of the Kasekela community of chimpanzees in the Gombe National Park on the shores of Lake Tanganyika had deliberately set out on a raiding expedition into the territory of the neighbouring Kahama community where they proceeded to wreak a terrible punishment on their unsuspecting victim. It was a pattern to be repeated over the ensuing months, until every one of the six males of the Kahama community had suffered the same fate. All died of their wounds.

News of this extraordinary event spread like wildfire through a shocked research community. Never before had such a thing been seen in any species of primate. And of all places to find it – in the hitherto peaceful world of chimpanzees. Granted, there had been odd outbursts of aggression among the males, but these had been isolated incidents – no more than the posturings of males on football terraces on a Saturday afternoon. The attacks on the Kahama males changed our perception of chimpanzees for ever. What made it seem worse was that the Kahama males had all originally been members of the Kasekela community, but had moved out to establish their own territory next door only a few years previously. They were all individually well known to their killers.

But why should the traumatic events of that day have seemed so shocking? After all, by human standards, the Gombe event was quite modest in scale. On 1 July 1916, the first day of the Battle of the Somme, General Haig's British army suffered 58,000 casualties (a third of them killed), and an estimated ten million civilians and combatants died during that great war-to-end-all-wars. During the Second World War, the Nazis managed to shoot, gas, burn or work to death some six million Jews and an equivalent number of Gypsies, Slavs, communists and other 'undesirables' in little more than half a decade. In a desperate attempt to subjugate the Belgian Congo during the closing years of the nineteenth century, King Leopold of Belgium's henchmen killed an estimated six million people. The Khmer Rouge accounted for two million in Cambodia during the 1970s, and the most recent tribulations in the Congo are estimated, at the time of writing, to have left nearly five million dead.

Recent history has played host to so many instances of genocidal conflict that they are too numerous to list. The massacre of Hindus and Moslems by each other during the months leading up to the partition of India in 1947, the massacre of a million Armenians by the Turks in 1917, Katanga, Biafra, the Congo, Angola, Uganda under Idi Amin, the Lebanon, Northern Ireland, Rwanda, the Congo again (by then renamed Zaire), Bosnia, Somalia, Kosovo, the Congo again (now back under its old name) . . . It is said that nearly 400,000 civilians were massacred and 80,000 women raped when the Chinese city of Nanking was sacked by the Japanese army in 1938.

And as we reach further back in time, so the unseemly list grows. The Balkans over and over again, first Slavs against Turks, then Turks against Slavs, then Slavs against each other, in an endless cycle of vendetta and retribution. Before that, there had been the Crusades, the Night of the Sicilian Vespers (when the Sicilians massacred their French Angevin masters in 1282), the 'harrying of the north' (of England) ordered by William the Conqueror in the aftermath of the Norman Conquest that resulted in tens of thousands dying of starvation or being put to the sword, the St Bryce's Day massacre of Viking men, women and children that occurred some seventy years earlier on the orders of the English king Ethelred the Unready.

Then there were the endless pogroms against the Jews during medieval times and beyond, the religious persecutions that dogged Europe in the aftermath of the Reformation and the Counter-Reformation, the chaos and misery of the Thirty Years War when rampaging armies from all over northern Europe occasionally slaughtered each other but mostly just wreaked havoc and vengeance on an impoverished and long-suffering peasantry. In the infamous Albigensian crusade of 1209, an army of 30,000 knights swept down from northern Europe at the behest of Pope Innocent III to slaughter the Cathar heretics of the Languedoc in southern France. Whole cities were razed, crops destroyed, property ransacked and stolen, populations put to the sword. In Béziers, 15,000 men, women and children were killed in a matter of days, many while seeking sanctuary in the churches. And after Béziers, the same fate befell Perpignan, then Narbonne, Carcassonne and Toulouse. No one was spared. As if that was not enough, 100,000 Protestant Huguenots were slaughtered by the French King's Catholic troops in one week in August 1572 in what became known as the Massacre of St Bartholomew. When news of that massacre reached Rome, cannons were fired and bells rung in celebration, and Pope Gregory XIII ordered a special commemorative medal to be struck.

And lest we think this merely a Christian European tradition, we need only consult the Bible's Book of Judges. Here, in battle after battle, tens of thousands were slaughtered in the ebb and flow of fortunes between one petty kingdom and another.

Although we should make allowance for the exaggerations of the victors in these histories, there is no suggestion of censure or pity in these writings. The men of Gilead, fighting under Jephthah, gave no quarter to the Ephraimites after they had defeated them. Instead, they tracked them down, flushing them out as they attempted to blend in among the local population by challenging them to pronounce the word *shibboleth* in the Gileadite manner and putting to the sword all who pronounced it *siboleth* with a sibilant 's' in the Ephraimite manner. '. . . And there fell at that time,' the writer of the Book of Judges calmly tells us in the sonorous tones of the King James Bible, 'of the Ephraimites forty and two thousand.' A few centuries later, and it was the turn of the victor's descendants to suffer the same fate at the hands of the Romans. In 135 AD, after a protracted and frustrating campaign, the Roman general Julius Severus destroyed fifty fortresses, laid waste to 985 villages and killed over half a million of their inhabitants in a determined effort to rid Palestine once and for all of the troublesome Simon Bar Kokhba and his guerrillas. A dozen or so centuries later in 1826, the British trader Henry Flynn was obliged to watch the 50,000-strong army of the notorious Zulu chief Shaka deal in similar vein with the Ndwandwe, the Zulus' most truculent enemy. In little more than 90 minutes, Flynn later recounted, the Zulus slaughtered 40,000 men, women and children.

Who are we to be so shocked at the behaviour of a handful of male chimpanzees?

But perhaps that is precisely why we were so shocked at the events that engulfed the Kahama males, for we had been led to believe, thanks mainly to a series of *National Geographic* films, that chimpanzees represented an innocent peaceful state of natural grace from which – at the moment Adam accepted the fruit of the Tree of Knowledge from Eve – we humans had so conspicuously fallen.

The Idyll in the Forest

Our earliest impressions of chimpanzees in the wild were almost idyllic. Chimpanzees, it seemed, lived in communities of 50–80 individuals. They roamed through the forests in search of fruits and berries, lolled in tree-top nests in the heat of the day, fished ingeniously for termites using carefully prepared grass stems and tended to their mischievous young in the best traditions of apple pie and motherhood. The picture was indeed alluring. There was gentle Flo, the grand matriarch, nurturing her boys Faben and Figan (later to be the dominant male of the community); tolerantly playing with Fifi her young daughter, born not long before Jane Goodall had arrived at Gombe, and later as proud grandmother watching over Fifi's own first baby, Freud. These were idyllic lazy days, relaxing on the forest floor as warm tropical breezes lapped fitfully at the treetops and insects buzzed contentedly among the flowers below. There were exciting new discoveries – the first evidence of termite-fishing, of tool use and tool-making, of Flo's maternal skills in patiently distracting the ever-fractious Fifi with tickling games.

To be sure, there was the occasional outburst from one rampant male or another seeking to challenge the community's hierarchy. These could be awesome displays that sent everyone in the vicinity, chimpanzees and human observers alike, racing for cover as bristling males swaggered about, crashing through undergrowth, deliberately running down innocent groups of females and young, throwing branches and saplings about with scant concern for whom they might hit. But peace would re-impose itself as quickly as the fracas had started, just as soon as male honour was satisfied and the exhausted protagonists had slumped into a tree fork to sulk haughtily. In human terms, these were no more than noisy youths on their motorbikes revving their way down the village High Street late on a Saturday night – gone and forgotten by church on Sunday morning.

Sometimes individuals exercised surprising ingenuity in these displays. Though smaller than most of the other males in the Gombe community at the time, Mike discovered that he could cow his rivals into submission by throwing around the empty kerosene drums left neatly stacked in Jane Goodall's camp. Not only did they make a satisfyingly resounding clatter when banged together, but they could do significant – albeit mostly superficial – damage when they hit other people. Mike won his way to the top through native ingenuity rather than brawn.

Meanwhile, a few hundred miles away high in the alpine forests of Rwanda's Virunga volcanoes, much the same kind of story could be told of the gorillas studied by Dian Fossey and others. In fact, the gorilla's reputation had been rescued by the field studies in Rwanda, first by the redoubtable American biologist George Schaller and later by Fossey and her students. Before this, the gorilla's reputation had been in the *King Kong* mould, thanks mainly to hunters' tales of being charged by irate males trying to defend their groups or to whispered stories about children (and occasionally women) stolen from the edges of villages and eaten (even raped) in the forest by these formidable apes. No doubt the male gorilla's size (an adult can weigh as much as 140 kilograms) and his unswerving determination to defend his harem of females against all comers were to blame for this. An adult male in full charge downhill can be a unnerving sight, especially when – as once happened to a colleague of mine – the by now increasingly alarmed gorilla suddenly realises that gravity and the steepness of the hill slope have conspired to deprive him of all capacity to stop.

But once serious field studies got under way in the 1960s, it seemed that the reality of gorilla life was a dull and exceedingly monotonous round of eating grimly unappetising herbs, interspersed with long stomach-rumbling sessions dozing in the warm midday sunshine, or gentle perambulations from one heath-covered hillside to another. Gorilla groups are small (typically less than ten individuals), tight-knit and seemingly well adjusted. The gentle giant merely added to the sense that ape society was more akin to the faded gentility of a seaside old peoples' home than anything else.

The only slightly disconcerting blip on the horizon was the orang-utan. Studied in Borneo and Sumatra by a string of keen young zoologists and anthropologists, the red ape – as the orang was soon dubbed – at first added to the growing image of pastoral bliss. Never very social at the best of times, orangs wandered the forests of their island home mostly alone, or at best accompanied by their dependent young. But males did sometimes engage in ferociously bloody fights when they chanced upon each other while on their peregrinations. And some male orangs, it soon became apparent, had a rather distressing penchant for rape. Although the larger males typically sat back and waited for females to come to them, the smaller-bodied males (in whom the females exhibited much less interest) relied on naked coercion to get their way. Being barely half the male's bodyweight, a female orang is no match for even a small male, and shocked observers could only sit and watch in disbelief as females fought unsuccessfully to avoid the inevitable.

Sex and a Singular Ape

And if ape life had seemed idyllic, this was only confirmed in spades when the bonobo was first studied in the wild. The Japanese field workers who undertook the first studies of wild bonobos in the Congo Basin in the 1970s were astonished and bemused to find that sex seemed to be the sole preoccupation of these unusual apes. They were, quite literally, at it all the time, and in all possible combinations – males with females, females with females, even occasionally males with males. Here was an almost unique example of sex being used outside the confines of mere reproduction to cement friendships, as a means of reducing tension, even as the currency for buying access to food sources. A male who found a particularly nice tree full of fruit would sit at the bottom of it and only allow females into it if they first had sex with him.

Nothing in the Japanese fieldworkers' extensive experience of common chimpanzees in the wild had prepared them for this. The bonobos' particular predilection for sex seemed quite unique in the animal kingdom, and made them seem almost human. That sense of humanness was only enhanced by the fact that they often had sex face to face the way humans do – all but unheard of in the rest of the animal kingdom – and, rather more disturbingly, by the fact that adults of both sexes often engaged in (mostly non-penetrative) sex with infants as young as a year or two in age.

First impressions of the bonobo's behaviour were that it differed radically from the common chimpanzee's. Most writers characterised it as a peaceful, sociable, loving, relaxed vegetarian with hardly a care in the world – a species quite unlike the aggressive, solitary, violent, fractious, meat-eating common chimpanzees of the Gombe and elsewhere. It is certainly true that bonobo society has a more peaceful air to it: the fraught scenes we have become used to at Gombe are less common. And it is unquestionably true that bonobos exploit the opportunities offered by sex to defuse tense social situations as well as to bond friendships with each other. In this, they resemble humans much more than they do common chimpanzees who rarely use sex other than for brute procreative purposes.

But, it seems that the bonobo's enviable reputation for peacefulness may not be quite as snow-white as it once seemed. Bonobo males can get just as nasty as common chimpanzee males when they really want to. Their problem has much more to do with the shadow of the females hanging over them. Bonobo females can get pretty feisty when it suits them – as the graphic observations by primatologist Amy Parish of a bonobo female biting off a male's penis in her rage at his overly persistent advances can only serve to remind us. In addition to the threat of physical assault, bonobo females benefit from a certain asymmetry in their favour. They have at their disposal that which the males most desperately want – sex. In a word, the males cannot afford to upset the females too much, because the females would be quite happy having sex with each other, leaving the males out in the more than merely proverbial cold.

On closer inspection, it seems that the bonobo's reputation for peacefulness may be based as much on an accident of circumstance as on any real difference in temperament. What marks bonobo society out from that of common chimpanzees is that they live in larger groups. And they live in larger groups because the forests they inhabit within the great bend of the Congo River in central Africa provide food sources that are very much larger than those in the more seasonal habitats typically occupied by chimpanzees. The supermarket facilities offered by the Congo forests (giant fig trees dripping with fruits) allow more animals to congregate in one place. In contrast, the more seasonal habitats of the chimpanzee force females (in particular) to spread out and forage in smaller groups, and often on their own. Once females are in sizeable groups, they are able to exert a great deal more influence on male behaviour, as studies of captive chimpanzees at Arnhem Zoo by Dutch primatologist Frans de Waal demonstrated. In this enclosed population, where the community was obliged to live together rather than disperse through the forest, the battles among the males for dominance were constrained by their need to keep the females on their side. Though not exactly acting as a calming influence on the males, the females did provide a counterweight that the wily old male named Yeoren did his best to exploit, often with considerable success.

The Odd Couple

All this might make us wonder whether we humans are really all that different in our behaviour from the chimpanzees, and especially the bonobo. The bonobo's rampant enthusiasm for sex under any circumstances is so uncannily close to what we see in humans that it is hard not to envisage a common origin for the two. But there does seem to be one difference between us humans and our brother apes: the rather peculiar phenomenon of pairbonding. Although female chimpanzees may form special relationships with individual males during the time that they are mating (and in common chimpanzees, this is pretty much confined to the time when the females are undergoing menstrual cycles), these relationships do not really persist – not even in bonobos.

Humans are clearly very different. Individual males and females become infatuated with each other – the peculiar state that we call 'falling in love'. This seems to involve an intense attraction to the partner, feelings of desperate longing when physically separated, a tendency for everything else in the world to lose significance and, last but not least, a curious tendency to endow the partner with all kinds of hopelessly unrealistic and inappropriate characteristics – to see them, as the saying goes, through rose-tinted spectacles.

This peculiar form of monogamous pairbonding is puzzling for two reasons. First, why on earth do we do it, when all the other great apes follow a much more promiscuous lifestyle?

Although monogamy is by no means unknown among animals, it is a rare form of social arrangement among mammals, being common only in the dog family (wolves, jackals, etc), where it is ubiquitous without exception, and among some (but not all) of the small African antelope. It does occur among primates, but only in a handful of very specific groups (the gibbons and the smaller South American monkeys). Second, it is even more unusual to have this kind of monogamous mating system as part of a large multi-male/multi-female community. In all other obligately monogamous mammals, from the entire dog family (wolves, jackals, coyotes, etc) to gibbons, the paired couple live alone on their own territory, accompanied only by their dependent young. Their response to neighbours is cautiously disdainful, and to intruding strangers positively hostile. That way their relationship is protected from the threat offered by having rivals in close proximity.

In contrast, humans have this peculiar habit of forming long-lasting pairs within large (sometimes very large) social groups. In the few other species that have a similar system (bee-eaters, a family of small wonderfully colourful African birds, being a case in point), this places a very intense pressure on the fidelity of the couple, not least because the female is subjected to constant sexual harassment by the other males in the colony. In these cases, the monogamous pairbond is a form of mate-guarding in which the male sticks like glue to the female in order to prevent her being picked up or harassed by other males.

Only two species of primates beside ourselves have a comparable kind of social system, both as it happens being indigenous to Ethiopia. One is the hamadryas baboon that inhabits the deserts of the north-east, whom we met earlier; the other is the gelada which is found only in the central Ethiopian highlands. In both species, males with their harems of two to ten females plus dependent young associate in large herds that, at least in the case of the gelada, can number upwards of several hundred animals. The persistence of the pairbond (or at least the male's monopoly over mating access to his females) seems to depend mainly on other males being inhibited from challenging harem-owners directly for their females. Males seem to go out of their way to avoid poaching each other's females – at least, most of the time. The 'most of the time' in this context seems to reflect the views of the females themselves, as was demonstrated by a series of inspired experiments on wild and captive hamadryas baboons carried out by the Swiss zoologist Hans Kummer and his students.

These studies revealed that this inhibition depends largely on cues given by the female: males will occasionally attempt to wrest a female away from her male, but only in those cases where the rival is *very* dominant over the paired male, or the female is showing a distinct lack of interest in her paired male (usually signalled by rather subtle cues such as a tendency not to stick quite so close to him or to pay quite so much attention to him as is normally the case).

My own field studies of gelada suggest that a male's hegemony over his little group of females (to whom he has exclusive sexual access) also depends on how the bachelor males read the runes of his females' interest in him. When a bachelor male attempts to take over a harem of females by displacing the current breeding male, the result is inevitably a bloody and prolonged fight with sabres fully drawn. A displaced male never gets the chance to breed again, so he is hardly enthusiastic about giving up his harem. But whether or not such an attack is successful depends entirely on the majority of the females being willing to desert their former male for the new one. In between bouts of fighting, the two males spend most of their time trying to interact with the females. The rival does so in order to persuade the females to signal their willingness to desert by grooming with him; the old male spends his time rushing from one female to another in a frantic attempt to groom each one, as though to make up for past lapses and lack of interest – the monkey equivalent of bringing a box of chocolates home after yet another late night out down the pub with the boys. The process is a simple democratic vote by the females: if more than half of them express a preference for the new male by grooming with him, the rest will follow in swift succession and the old male can do nothing except retire as gracefully as he can manage. Further resistance is utterly pointless.

One particularly revealing example of the importance of the females' preferences occurred during one such takeover. In most cases, successful takeovers involve large harems with many females. Such units often have an additional young male (known as a 'follower') associated with them. Having entered the unit in a very submissive capacity, the follower male spends a year or two slowly building up a relationship with one of the more peripheral females, with whom he will eventually leave to establish a harem of his own. After the takeover of one such unit, the successful male set about cementing his relationships with each of the females in the unit in turn by grooming and mating with them. Eventually, he got around to the female bonded to the follower. He adopted the approach of launching an attack on the follower, much as he had just done with the harem holder. Being younger and a good deal smaller than the new male, the follower responded by throwing himself on the ground and screaming in abject fear every time the male came near him. After all, he had just seen what had happened to the old harem male at the hands of this thug. However, every time this happened, the follower's bonded female came rushing across and, despite being half the male's size, launched herself at the new male, spitting fire and brimstone. The message was very clearly: 'Get lost – I'm spoken for!' After several attempts to split the couple apart, the new male eventually gave up and made do with the five females he'd gained from the takeover, leaving the sixth to the follower. What is interesting about this is the commitment that the female seemed to be showing to 'her' male. Despite the immediate capitulation by the follower and the obvious physical dominance of the new male, she was having none of it.

These observations highlight the fact that pairbonds of this kind are a risky business: both internal conflicts of interest and external pressures (the sudden availability of more interesting rivals) combine to put considerable pressure on a relationship when these exist within a larger social grouping. It also seems that, at least in these primate examples, the cues that rivals use when deciding whether to risk trying to split up a pair are those that reflect the intensity of the pair's interest in each other, and particularly the female's interest in her bonded male.

The risks of desertion that we run – and the opportunity cost that we incur in terms of losing both what we have already invested and the benefits of what we anticipate gaining from the relationship in the future – clearly weigh heavily even in our case. In humans, sexual jealousy and fear of desertion are the principal causes of within-pair conflict, as has been amply demonstrated by studies of household homicide and domestic violence in Canada and Europe.

In most cases of such within-pair conflict, the man is the aggressor and the woman the victim. But women are not entirely exempt. As British psychologist Ann Campbell's detailed studies of violence in women in both the USA and Britain have revealed, the risk of losing a preferred mate can precipitate concerted attempts to scare off the opposition. The main difference, it seems, is that men threatened with desertion typically direct their attempts to protect their relationship against the offending spouse, whereas women tend to direct theirs against the rival. This may well be a consequence of the fact that, for either sex, attacks on the offending male are more likely to lead to a damaging escalation of physical aggression than are attacks on the female.

Although women in such situations typically react less violently in physical terms than men, their responses can be just as damaging to the victim – and, in some cases perhaps, more cruel. The traditional weapons of women seeking to protect their reproductive interests are to abuse opponents verbally, casting aspersions on their character, issuing threats of physical consequences in the future, verbal bullying and psychological warfare. It may, of course, be that the tendency for women to be physically more controlled on these occasions is simply a reflection of socialisation, perhaps a consequence of men selecting more feminine women who do not exhibit traits that are more characteristic of men. Even though casual observation on the street suggests that, in contemporary British and American society at least, gratuitous assaults by young women may be increasing in frequency, it seems nonetheless true that women the world over are less physically violent than men.

Mechanisms of this kind, that are designed to protect pair-bonds as much as anything else, bear a striking resemblance to the kinds of strategies used by hamadryas baboon males to enforce their control over their females. In this species, another male drifting too close to a female does not incur the wrath of the female's bonded male; instead, it is the female who bears the brunt of her male's concern. It is she that is attacked by the male, often with vicious bites to the neck. It is a process of intense negative conditioning: females very quickly learn not to stray too far from their males.

That this is a learned response on the females' part is indicated by two telling observations. One is that, as hamadryas males age, they become less assiduous in their herding of their females; as a result, the females become less concerned to stay close to their males, and are more easily prised away by passing rivals. The second is what happened when Kummer released two hamadryas females into a group of common baboons (whose social system is less structured and more promiscuous than that of the hamadryas, with females being rather more independent spirits). Although the females initially attached themselves to a male, they very soon learned that drifting away from him did not incur a penalty, and after a few weeks they started to behave like perfectly normal independent-minded common baboon females. Conversely, when he released a common baboon female into a hamadryas group, she showed the reverse behaviour: quickly adopted by a bachelor male, she rapidly learned to behave like a good hamadryas female, keeping attentively close to his heels whenever he moved.

Violence is, however, a weapon of last resort, because it can in the end backfire. At best, it may increase the victim's desire to escape; at worst, it will make the victim an unenthusiastic mate, reluctant to offer those benefits of pairbonding that we so crave, leading almost inevitably to a viciously destructive spiral of violence. Instead, we humans often use a whole arsenal of more subtle tactics to dissuade our partners from abandoning us in favour of someone more enticing. All are inevitably underpinned by the same emotion – jealousy triggered by fear of loss – but some at least rely on psychological persuasion rather than brute violence.

These behaviours can include psychological manipulation (appeals – tearful or otherwise – to the mate's better nature) or forms of entrapment (notably, instigating a conception), but they can also include the subtle removal of the mate from the situations where the pairbond is likely to be put at risk. The latter include the use of purdah and other forms of partial seclusion like the wearing of the *burkah* that are familiar examples of the way men seek to restrict women's access to potential rivals in the Middle East. In classical China, footbinding was carried out during childhood and so deformed the woman's feet that she was unable to walk more than a few yards on her own. It functioned as an honest signal of the woman's faithfulness, since she was simply unable to steal off to lovers' trysts in the dead of night. (Footbinding, incidentally, was practised mainly by the wealthier classes, who had more to lose if their daughters were caught in flagrante delicto.) Or it may involve a heightened level of attention to the mate – the Scheherazade effect, whereby an individual seeks to keep its mate entertained and interested. Either way, the aim is to prevent the partner from interacting with rivals.

The Brain Strikes Back

The human brain weighs around 1.2 kilograms, compared to a mere 400 grams for the chimpanzee's. Even taking the difference in body size into account, the human brain is still around double the size we would expect for an ape of equivalent body size. It is something like six times larger than we would expect for a mammal of our body size. The cost of rearing an offspring with a brain of this size is enormous, and accounts for much of what differentiates our biology from that of other animals. Not only has the human brain played havoc with our anatomy in order to make gestation and birth possible, but it imposes monstrous demands on its progenitors after its arrival. It requires a long period of lactation and, beyond that, many years of nurturing and socialisation to turn what begins as little more than a wet lump into a half-decent human being.

The slow rate at which brain tissue is laid down and matured means that the period of direct parental investment (pregnancy and lactation) is greatly drawn out in anthropoid primates (the monkeys and apes). Humans simply represent one extreme of the distribution. Pregnancy amounts to the conventional nine months, but the human brain requires a further year before it completes its physical growth, plus an additional four years or so before the child is mature enough to survive on its own. Not merely does the growing brain need nurturance and care, but there is also the threat of many extrinsic dangers. Susceptibility to childhood diseases is particularly severe during the first five years and was the major cause of death in premodern societies.*

The costs of rearing children are so high in humans that it is difficult for a single parent to provide sufficient care. Orphans and the offspring of single parents bear a heavy cost in terms of higher-than-average mortality even in our own more enlightened and economically better-off societies. In earlier societies, the costs of single-handed rearing commonly led to high levels of abandonment and infanticide. During the eighteenth century, for example, the frequency of infant abandonment in the city of Limoges in France correlated with the level of hardship (as indexed by the price of barley) in any given year. The frequency of infanticide and abandonment of children by the poor (especially among single women, widows and women living alone) in Victorian England reached such levels that it came

* Unfortunately, this has resulted in some curious misunderstandings about how long people lived in the past. The much vaunted low average lifespan of our Victorian forebears (often said to be between thirty-five and forty-five years, compared to between seventy and eighty in contemporary western populations) was in fact largely a consequence of very high childhood mortality within the first five years of life. Even a cursory glance around your local churchyard will demonstrate that those of our Victorian forebears who survived childhood more often than not died at commendably ripe old ages in their sixties and seventies. At least two of the dozen or so eighteenth- and nineteenth-century male members of my paternal lineage – all crofters or estate workers from the north-east of Scotland – survived into their nineties, and most made it into their seventies.

to occupy a great deal of parliamentary time, and eventually culminated in the 1922 Infanticide Act.

In many ways, the value of paternal support from the mother's point of view is attested to by the extent to which polygamy is practised. The issue, however, is not so much that it is widely permitted, but rather who does it. Although polygamy is probably the most common marriage system in terms of the number of societies permitting it, in practice it is less common as a *living* arrangement than monogamy, even among those societies that approve of it. This is because women in these societies will only accept being the second or third wife of an already-married man if he is wealthy enough to support an additional wife and her children. The evidence from many studies is quite uncompromising: on average, women who marry relatively wealthy men have much higher success rates in rearing children than those who marry poorer men, when number of wives is held constant. This was even true in eighteenth- and nineteenth-century rural Europe, as we can see very clearly from the church registers of births and deaths: child mortality rates declined as landholding increased. The reason is very simple: the wife of a wealthier man has access to more resources that can be invested into her offspring, in the form either of food or of medical help (whether this is of the traditional or the western kind) when they are ill.

The costs of rearing, and the consequential need to ensure that a spouse is attracted and retained long enough to ensure successful rearing, are thus likely to have been the principal factor favouring the evolution of pairbonding in our species. Something is needed to ensure that the couple stick together for long enough to rear the current offspring. Such devices are by no means unusual in nature. The maternal instinct that appears as if from nowhere after women have given birth plays a similar role in tiding us over the period when the human infant is utterly unresponsive but in desperate need of care and attention. Later, the infant will be able to trigger those responses from its parents for itself by smiling, gurgling and using all the wiles that only babies have to elicit that 'Aah' response in adults. But in those first few months of life, the human infant (unlike its primate counterpart) is pretty much an inert lump. Something is needed to induce parental care in those fraught early days if the infant is to survive. The maternal instinct seems to be triggered as part of the hormonal restructuring that is brought about by the birth process itself. It is a remarkable piece of evolutionary engineering. And it provides a nice example of how hormonal and emotional processes can interact with cognition to achieve essential biological objectives.

All this would seem to explain fairly convincingly why women should be subject to this phenomenon of falling in love. But why should this be relevant to men? One answer, of course, is that men are (in evolutionary terms) forced into the business of helping to rear the offspring, even if in many (but by no means all) cases they do it simply by providing resources for the mother to use. However, they have to be willing to incur this burden in each case, even though many might argue that they would be better off spending their time down at the pub (or the appropriate equivalent in the society to which they belong). In part, of course, they are under the same kind of pressure as their spouses: if rearing takes more than one parent, then men who do not take part (directly, or indirectly through creating wealth) are not going to be as successful at leaving descendants as men who do. So they, too, need something to focus them on the job. Exploiting the same physiological and hormonal mechanisms that produce this effect in women is an easy option, particularly as the mechanism is already there in one sex. But there is another consideration that might add to the pressure on males: the risk of infanticide. If the male abandons his spouse in favour of the pub, or perhaps another woman, he risks another male taking over his spouse and killing the offspring in the interests of being able to breed with her.

It is a curious fact that the topic of infanticidal males has generated more hot air and nonsense than almost any other topic in evolutionary biology. Sadly, much of it has been the consequence of misunderstanding about what exactly is involved. This is unfortunate, because it can only serve to obfuscate and confuse. So, to avoid further confusion, let me pause for a moment to spell the issues out as clearly as I can.

In mammals in general, the mother's return to reproductive condition is largely determined by the length of time that the baby needs to be fed by the mother before it can be weaned. Generally speaking, weaning is dictated by when the brain completes its growth and the infant can manage alone. Because the rate at which brain tissue is laid down is very slow, and seems to occur at a constant rate in all mammals, primates' unusually large brains mean that they are obliged to put up with extended periods of postnatal amenorrhoea and long intervals between successive births. In apes, for example, females give birth only about once every 5–8 years. Humans are somewhat exceptional as apes go because they have much shorter inter-birth intervals than is typical of the other apes. However, even in humans, the average inter-birth interval is in the order of four years in traditional societies. Only in post-industrial societies where bottle-feeding and early weaning to commercial baby foods is the norm can women give birth at intervals as short as 12–18 months.

The costs of trying to hurry the process up are all too apparent in the demographic statistics of both humans and other primates. In rural Germany during the eighteenth and nineteenth centuries, for example, the likelihood of the second-born infant surviving its first year of life was directly related to how long after its older sibling it was born. Too soon, and its chances of dying rose dramatically because the mother could not cope with the costs of rearing two infants simultaneously. Much the same is true of the San foragers of south-west Africa. San mothers try to engineer a four-year birth interval by imposing very strict cultural prohibitions on intercourse during lactation; infants that are born at much shorter intervals have much reduced chances of surviving. Although women in all societies may deliberately try to engineer optimal birth spacing, it is not entirely a matter within their control. Even though the mother will inevitably do her best to protect her offspring, her body will draw the line at putting itself at risk and will start to shut down milk production if the load on her is so great that she begins to lose too much body fat and muscle. Being inevitably the weaker and more dependent, it is the second infant that is most likely to suffer.

The issue for all primate males, human and non-human, is that a male who inherits a pregnant or lactating female from another male will be unable to reproduce with her until weaning has been completed, and this could be many years away. In those societies where males compete to 'own' groups of reproductive females, a male's reproductive tenure is likely to be much shorter than his lifespan because he typically acquires a harem only relatively late in life and then risks being ousted or killed by other males well before age has dimmed his reproductive powers. If the average tenure is less than the typical length of a reproductive cycle (the interval between one birth and the next), males who do not behave infanticidally will fail to reproduce. In contrast, males who kill existing infants will benefit because the mothers will return to menstrual cycling almost immediately. This is because, in all mammals from red deer to humans, it is the infant's suckling action that prevents the return to menstrual cycling. So long as the baby suckles at a rate in excess of about one bout every four hours, the action of suckling interferes with the female's hormonal system and prevents the build-up of the gonadotrophins that manage the menstrual cycle. This is why women who bottle-feed begin cycling again sooner than women who breastfeed.

Because male mammals who acquire a pregnant or lactating mate will be unable to reproduce for some time, infanticide has been widely documented in all the mammals, but especially so in primates. Among gorillas, for example, it has been estimated that around 30 per cent of all babies born are killed by males. In some Amazonian tribal societies, as many as 45 per cent of the babies born fail to survive to five years of age, in large measure due to infanticide. Among the Ache of Paraguay, the men are quite explicit: they simply cannot afford to rear another man's offspring when they take over his former spouse and, if he is not around to protect it, they will simply do away with it, the mother's protestations notwithstanding. For them it is not a big moral issue, simply a matter of the practicalities of survival – and, of course, reproductive access to the mother. And the mother's interests are, in the long run, best served by developing a functional relationship with her new male as soon as possible. This can strike us as harsh and uncaring, but our attitudes to children in the west are much affected by the dramatic reduction in family size that has occurred within the last century or so. With typically only two children to play with now, we need to be a great deal more committed to each one. It has not necessarily always been thus.

These are, of course, extreme examples and most societies or species do not experience rates of infanticide even close to these levels. If they did so on a regular basis, they would soon become extinct. However, the fact is there, bubbling beneath the otherwise calm exterior of society. Unfortunately, our more squeamish attitudes in the west have inclined many to try to sweep the phenomenon under the carpet and deny its existence rather than trying to understand and explain it. So it is important to be clear that the issue from an evolutionary point of view is not that most babies survive in most societies and species, but rather that infanticide occurs at all. That infanticide exists and can, in some cases, reach quite serious proportions, turns out to have very significant implications for many aspects of behaviour. And one of those is pairbonding. Providing protection for one's young offspring becomes a crucial service that males can provide. For males, protecting your investment in your own genetic future provides an additional important selection pressure favouring the evolution of pairbonding in our species. Detailed analyses, in some cases involving mathematical models, have shown that infanticide is the most likely explanation for pairbonding in both gibbons and gorillas.

Pairbonding is not, of course, necessarily a permanent state in humans – or any other species, if it comes to that. There is ample evidence that the emotional bonding that occurs between couples can and does break down with time, with partnerships disbanding as a result. Among the Ache hunter-gatherers of Paraguay, adults average around a dozen partners over the course of a lifetime, with each partnership lasting from a few months to several years – a situation that may not be untypical of human hunter-gatherer societies where formal marriage arrangements do not exist, as well as being one which we in the west are rapidly approaching now that formal marriage is no longer considered absolutely essential.*

However, infanticide is by no means the whole story: humans, uniquely among the primates, may require the hard work of two carers for successful rearing. After taking the dif-

---

* The very large number of spouses observed in these traditional societies reflects not only the natural dissolution of pairbonds over time, but also the fact that spouses may often die of disease or injury.

ference in body size into account, the energy costs of producing an infant are about 10 per cent higher for humans than for chimpanzees due to our much larger brains. In addition, human infants are much less mobile than those of other primates, and thus have to be carried by the mother for much longer. This partly reflects the long period of dependency in human infants, but it also partly reflects the fact that human infants are born about twelve months premature compared to the babies of all other primates. They just need a massive investment, and it requires two adults with a personal genetic commitment to the infant (or in the case of the male, what he is *persuaded* is a genetic commitment) to ensure that sufficient energy and care is channelled the baby's way. In traditional hunter-gatherer societies, the need for one member of the rearing group to be free to forage or hunt probably made rearing units consisting solely of women unfeasible.

Other arrangements, however, can and do work, although much probably depends on the exigencies of economic circumstance. When men are able to monopolise resources (such as land or other forms of movable wealth like stock animals) that women can invest in rearing offspring, then women may be willing to marry polygamously. Women seem to be prepared to put up with such marital arrangements in order to gain access to these resources precisely because these resources have a huge impact on the survival and future of their offspring. Wealth still does so, even in our own post-industrial societies: the better-off have lower rates of childhood illness and death, and are better able to place their offspring in society as adults (in terms of providing them with the educational and social opportunities that continue to underpin both social and reproductive success in modern societies).

The overriding importance of resources in the business of successful rearing is reinforced by a rather unusual form of family arrangement: polyandry among the Tibetans. In this case, the need to prevent farms being subdivided in successive generations is solved by ensuring that there is only one marriage in each generation (with all the sons of the family marrying one wife and contributing through their labour to the rearing of the 'communal' offspring). Farms that are subdivided between inheriting offspring in successive generations too many times very quickly become too small to support a family – a problem that was recognised very early on by the European landed gentry, and solved by them in the thirteenth century by switching from partible inheritance (equal inheritance by all the offspring) to primogeniture (inheritance of the whole estate by the oldest offspring).* Such solutions are never cost-free, however. Marital tensions in polyandrous Tibetan households can be intense. It is very difficult for wives in polyandrous households (and, if it comes to that, husbands in polygamous households) to be wholly even-handed in their behaviour towards their spouses. In the Tibetan case, younger husbands often leave if they can afford to establish themselves in monogamous relationships elsewhere, and the women seem to suffer unusually intense periods of psychological breakdown that may be designed to alleviate the stresses of balancing the interests and demands of several husbands.

In the polygamous households such as those of the Mormons or many of the African Bantu tribes, the men also often complain that coping with several wives can be very demanding. Polygamous marriages often lead to levels of stress between co-wives, with the result that polygamous marriages are rarely as fertile (on a per capita basis) as monogamous marriages in

* Interestingly, in Germany, ultimogeniture (inheritance by the youngest offspring) became the norm among farming communities. This probably helped to minimise the amount of disruption that occurred by making the interval between successive inheritances as long as possible.

the same society, probably because the stress disrupts the women's menstrual hormones such that they have relatively more cycles in which ovulation does not occur. In many Bantu societies, each wife has her own house or hut which the husband visits for a few days at a time in strict rotation. For co-wives to live together under the same roof is common only when they are closely related – usually sisters.

So, while we humans share a number of features of our behaviour with other primates (especially our ape cousins), we also exhibit a number of key differences. Most of these seem to be fall-out from our big brains. We share both the good and the bad habits of our primate (and especially ape) cousins, but it would be only a slight exaggeration to say that, on both counts, we simply do them on a grander scale. We share behavioural traits with them simply by virtue of our common ancestry. But, we, as much as they, have undergone some six million years of separate evolution since we last shared a common ancestor. In those six million years, we have had to add new traits and adjust and tweak old ones to accommodate the twists and turns of other aspects of evolutionary history. There are questions we can ask about why those particular twists and turns occurred, but the fact is that, even if we cannot at this stage offer more than shallow speculations, those twists and turns occurred, and we live with the consequences now. Like all our ape cousins, we have our endearing side, and we have our utterly disreputable side.

Before concluding this chapter, however, it might be as well if I clarify one last important point. Evolutionary explanations of human behaviour continue to have a bad press. Many contemporary critics seem to think that evolutionary explanations are explanations couched in terms of the genetic determination of behaviour. Unfortunately, those who espouse such views inadvertently confuse two very different kinds of explanations that biologists sometimes give. Biologists conventionally draw a clear distinction between questions about function (*why* something happens, the purpose it serves in the individual's life), mechanisms (*what* bodily machinery, including motivational systems, produces the effect), ontogeny (*how* the effect is produced during the processes of development) and history (*when* it came about in the species' evolutionary history). These questions (now known as 'Tinbergen's Four Whys' after the Nobel Laureate ethologist Niko Tinbergen) are quite independent of each other. Confusing them leads to mistakes that can result in seriously misleading conclusions.

The most common confusion (and the one we need to dispose of here) is that between function and ontogeny – the goal the animal is trying to achieve (in biological analyses, this goal is always genetic fitness – its genetic contribution to future generations) and the reason why it can behave in that way (which is always some combination of its genetic inheritance, environmental effects and experiential learning, including, in humans, cultural transmission). The crucial difference here is between the (developmental) causes of behaviour and its evolutionary consequences. That the goal of behaviour is maximising genetic fitness does not mean to say that the origins of that behaviour (in terms of its development in the individual) are genetic. The capacity to be able to make the decision to behave in a certain way may be genetic, but that does not mean that the decision to act in a particular way is itself genetically determined. It is the capacity (to all intents and purposes, the brain) that allows the organism to evaluate the costs and benefits of alternative behavioural possibilities, and so to make its choice on the basis of a free decision after weighing up its options.

This is a particularly important issue in the context of infanticide, an issue that has been much misunderstood. The fact that infanticide is an evolutionarily adaptive strategy does not mean that every male has to behave infanticidally: if that were the case, extinction would rapidly beckon. The biological issue, rather, is that males have the capacity to behave infanticidally: whether they act infanticidally or not depends on circumstances in each particular case. In biology, everything is context-dependent. Everything depends on the balance of costs and benefits on a number of dimensions that affect the male's future social and reproductive opportunities. Killing a female's offspring in order to mate with her is hardly the best courtship strategy on offer, so males will, as a rule, generally be a little more circumspect in how they behave. However, the option is always on the table as a tactic, and the risk is always greatly increased if the male can act with impunity. The female, too, makes the same calculations, and might conclude that acquiescing in the male's behaviour is the better long-term option.

The bottom line here is that having one's behaviour determined wholly by one's genes may be OK for an amoeba, but it simply will not do for anything much more advanced than that. Indeed, it would be evolutionary suicide, because most animals of any size cannot reproduce fast enough to track changes in their environment by changes in gene frequency alone: they have to use flexible behaviour, based on learning, to hold them in the game of life long enough to allow the biological changes to take place. The real world is statistical and uncertain, and large, slowly reproducing organisms like mammals and birds need to be able to respond accordingly.

A second important point that we need to remember is that, in biological (or evolutionary) terms, we are always caught between conflicting interests. Evolution favours those traits that result in the largest number of future descendants, but we have many alternative ways of achieving that goal. We can reproduce ourselves, or we can forgo breeding to help our rela-

tives reproduce more effectively; and if we do choose to reproduce, we can have many offspring and leave them to fend for
themselves, or we can have few offspring and dote lovingly on
each. In either case, there will always be some point at which the
two alternatives are equally profitable in terms of the numbers
of descendants they yield. This is because, although there are
universal rules in biology, these rules are applied in circumstances that can vary enormously from one individual to
another or, for any given individual, from one period of its life
to another. Because the optimal strategy depends on the balance of the costs and benefits created by individual circumstances, outcomes will always be contingent. There is never any
'right' way to behave in any absolute sense – merely choices
between alternatives that will be more or less profitable (when
measured in terms of consequences for genetic fitness) for a
particular individual in a particular set of circumstances.

5 So Sweetly Sung in Tune

For a moment, he had lost himself in reverie. It must have been the rhythmic movements of the two women working at their skin that had triggered the recollection. But in his mind's eye, he had conjured up a vision of the camp a dozen nights before. The firelight cast a warm glow through the dark of the evening, the flames flickering on the line of men as they stamped rhythmically in a ring around the fire. As they circled round, rocking gracefully in unison from side to side, they sang and hummed an ancient song. Beyond them, on the edge of the circle of light, the women stood in a cluster, clapping and beating the rhythm on hollow logs and tortoise shells, bursting into shrill song at intervals, encouraging the men to circle faster by stepping up the beat imperceptibly on each round.

For the men, the effort was beginning to tell. Sweat was streaming down their bearded faces, dripping onto their naked chests and thighs. The singing grew more intense, the stamping more forceful, the body movements, as they dipped their shoulders first one way and then the other as they went round the fire, more exaggerated. The deep hum of the men's voices alternated with the soaring ululation of the women, resonating in some mysterious inner cavern in the listeners' minds to create a sense of urgency and tension that made standing still seem an impossibility.

It must have been then that the old woman had detached herself from the group beyond the firelit circle to join the men, sliding almost unnoticed into the pulsating line of bodies, her own body, infused with a renewed life force, swaying in perfect time to the music and the men's rhythmic flow. One by one, other women and the children joined them, until almost the whole band was swaying and shuffling round the fire's embers. Only a handful of women, hampered by young babies at their feet or cradled in their laps, remained to provide the chorus whose throbbing rhythms and bursts of song seemed now to provide the very engine for the dancers.

He always marvelled how the women knew when the moment had come for them to join the circle of the men. There was no signal, but it was usually one of the older women who initiated it, eyes glazed and ancient body shimmying in perfect time with the rhythm. He was never aware that there had been any change of step, or some especially shrill ululation, to signal that moment. It just seemed to happen, and always at the point where his mind seemed to give way to a sense of lightness, of euphoria, a feeling of great communality with the others, in which differences gave way to bonhomie, grievances to the mellowness of fellow-feeling. It was hard then to maintain anger or distrust towards others. The ease with which these feelings dissolved and vanished never ceased to amaze him. They did not begin with this as an intended aim, these dances, but the outcome was always the same: afterwards, the group felt more strongly bonded for a while, everyone more generous towards others, more willing to share or help out, more cheerful.

Whatever else may be the case, humans differ from all other species of animals – including our ape cousins – in one obvious respect: language. There are some 4000 species of mammals, and around 10,000 species of birds (just to consider the so-called 'higher' vertebrates), yet we are the only species that has this particular faculty. To be sure, all these other species communicate with each other, sometimes in surprisingly sophisticated ways. But none of their communication systems can match human language for sheer flexibility and information-carrying capacity. Bees can tell each other in which direction and how far away they have found a nectar source, but they cannot comment on the merits of this nectar source relative to the one they visited yesterday. Nor can they comment on the flying conditions today or how atrociously the queen bee behaved towards her drones last week . . . or what they should all do next year when the season for founding new colonies comes around. And in the final analysis, neither they nor any other animals have ever used their communication systems to produce a literature of any kind.

Yet human language can do all these things, and more. It allows us to engage in such exotic projects as constructing rockets to send people to the moon and beyond, something that would be genuinely impossible without the cooperation of many individuals (each engaged in a carefully orchestrated set of very complex tasks) and the accumulated knowledge of many generations of individual scientists (whose individual, sometimes quite specific, discoveries were passed on to successive generations only because of language). For both of these, language is crucial. Language is essential if we are to coordinate the activities of so many individuals scattered in so many different places. Without the knowledge handed down from one generation to the next, it would not have been possible for all those individual scientists and engineers who eventually put together the first rocket to the moon to develop the technology needed to do it.

Why, then, do humans have this unique capacity? Indeed, why is it unique to humans? How may this remarkable capacity for language in fact be intimately tied up with several other equally unique, but often ignored, aspects of human behaviour, namely laughter and music?

Why did Language Evolve?

Language evolved to enable humans to exchange information. Beyond that, however, there has been a general assumption that what was exchanged must be information about the environment or about how to do things – 'There are bison down at the lake' or 'This is how you make a handaxe'. Language as a facilitator for the exchange of technical knowledge.

The problem with this view is not so much that it could not have been thus, but rather that it does not make a great deal of sense in the light of how people behave when engaged in these kinds of activities. To be sure, we do tell each other about the bison down by the lake; but, when we later set off to hunt them, we typically do so in silence. Hunting parties are often small (rarely more than six individuals in any modern hunter-gatherer, quite often a single individual working alone) and hunting itself is typically a silent activity. The whole idea, after all, is not to spook one's prey by engaging in a continuous chatter about how nice the day is and how much you look forward to enjoying a good rump steak at the end of a successful hunt. Likewise, when we teach each other how to make stone tools or clay pots, conversation tends to be limited to such profound utterances as 'Just watch what I am doing . . .' We learn most everyday tasks by trying them out for ourselves, not by verbal instruction. In these cases, language seems to be little more than a mechanism for grabbing attention: 'Pay attention!' is about all you really need to say. The complexities and subtleties of grammar seem superfluous. Something else must be afoot.

That something is hinted at in what people actually do talk

about. What occupies most of our conversation time is social topics – what we like and dislike, what others were up to yesterday, how So-and-So behaved, what the children have been up to, what is planned for tomorrow's outing, how to handle a difficult social situation at home or at work. Such topics account for two-thirds of our total conversation time. All the rest – politics, culture, technical topics, music, even sport – between them make up a mere third. This is not, of course, to say that all conversations have this breakdown, or even that all individuals show the same proportions. They clearly do not. It will surprise no one if I say that women's conversations tend to have a slightly higher proportion of social topics (though not all that much higher – perhaps three-quarters instead of two-thirds) while men make up the difference by spending more time talking about sport and . . . well, technical 'how to' topics.

Of course, we do engage in technical conversations about our work or the best way to find free downloads from the Web, or even the grammatical structure of Bob Dylan songs. Indeed, some of us can spend hours immersed in the joys of such topics. But, the fact is that, except for a handful of real aficionados, most of us tire rather quickly of these topics. There is nothing worse than the cocktail party bore who wants to tell you everything he knows about chess moves or what he found on the Internet yesterday. It tends to produce a response something along the lines of: 'Lovely talking to you, but I *must* have a word with Jemima over there . . .' or 'Well, I think it's time I filled my glass again . . .'

But if the conversation should turn to a discussion about someone we know, or to our individual experiences of life, that which ought to be tediously boring – or so the conventional argument seems to assume – will do to keep us happily at it hour after hour. That the social domain should feature so prominently in our conversations cannot be entirely accidental.

Nor can it be written off as idle chatter filling up the time between rare but important conversations. Nature is rarely so profligate: she does not evolve features that are used purposefully only very occasionally and are left with the engine idly spinning away in neutral the rest of the time. More typically, when biological features are used only intermittently, they tend to be used when needed and then tucked away out of sight. Among some species of animals (including, I might add, some primates), entire organs like the testes – normally of such perennial concern and interest to males – regress and all but disappear outside the mating system when they are no longer needed. The whole machinery of reproduction in women, from womb to breast, undergoes sudden and rapid enlargement for the few months that it is needed, afterwards reverting (more or less) to how it was before. No, our fascination with the social world is not an epiphenomenon, a trivial by-product of something more important. It *is* the point, the *whole* point and nothing but the point. Like it or not, this is precisely what we have to explain.

So why might language as a social phenomenon have evolved? The answer, to cut a long story short, lies in the social brain hypothesis. We saw, in Chapter 3, that there is a close relationship between neocortex size and group size in primates. The group size that this relationship predicts for humans is around 150, and groups of this size seem to be a common feature of human social systems the world over. The principal mechanism that primates use to bond their groups is social grooming. We do not really understand how grooming bonds groups, but the fact is that the amount of time spent grooming other members of the group by different species of monkeys and apes is related directly to their typical group size. The bigger the group, the more time the animals spend grooming each other. If we use the relationship between group size and grooming time to predict how much time humans would need to spend grooming if we bonded our groups in the same time-honoured way as other monkeys and apes, grooming would have to occupy something in excess of 40 per cent of our active day.

That sounds like a terrific idea (and I'll come back to why it would be so great in a moment), but for any organism that has to earn its living in the real world, that would be a wholly impractical proposition. The business of finding food is very time consuming. So much so, in fact, that no species of monkey or ape devotes more than 20 per cent of its waking day to social interaction (most of which is, of course, grooming). It is simply a matter of time budgets: there are only so many hours in the day and most of those have to be given over to the business of finding food. They just cannot afford to devote more than 20 per cent of their time to social activities. If they do, they won't balance their energy budgets.

It seems that we humans cannot improve on this amount of time. Samples of how much waking time is devoted to social interaction (most of which is, of course, conversation) in societies ranging from modern Europeans to traditional farmers in New Guinea and pastoralists in East Africa yield a figure of almost exactly 20 per cent. It seems that, even though we use language rather than grooming as our bonding mechanism, we cannot create time to make more available for social interaction. Rather, we seem to push the primate limit as far as it will go and simply make better use of our time.

Language is what allows us to do this and it does it in a number of different ways. The simplest is that it allows us to interact with several people at the same time. If conversation is basically a form of social grooming, then language allows us to groom with several individuals simultaneously. However, there is a limit on the extent to which we can do this. If conversation is in free-for-all mode, the upper limit appears to be three others. If a conversation involves more than four people (one speaker and three listeners), it will split up into two separate conversations within at most half a minute, as a casual glance around any party will show.

One reason for this seems to be that when there are more than four people in a conversation, the distance across the circle becomes too great for all the speech sounds to be heard distinctly against background noise. The effort of trying to hear what is being said becomes too much: we're not quite sure what has been said, so we hesitate to chip in to the conversation. Instead, we turn to the person next to us and start chatting to them. In addition, the opportunity that each of us has to hold the floor (and hence contribute to the conversation) declines rapidly as the number of people involved increases. In a dyad, each member can expect to spend around 50 per cent of the time talking, but in a group of five each person can only expect to talk for 20 per cent of the time – so being involved in the conversation becomes less and less worthwhile (unless of course your principal interest in life is just to be a listener).

The only way we can increase the number of listeners beyond three without leading to several simultaneous conversations is to impose draconian rules on who can do what. We either have to have a chairman who dictates who can speak when, or we have to make a formal arrangement that ensures that only one person is allowed to speak while everyone else agrees to maintain a dutiful and respectful silence (as happens at a lecture or a sermon). Of course, in the latter case, many of the listeners probably drift off to sleep (metaphorically, even if not always literally) because they lose focus and concentration. Paying attention for a long period is hard work. In a sense, you might almost say that speaking is a form of rest taken to alleviate the really hard work of listening.

What makes an everyday conversation work is the fact that we *interact*. Each of us needs to have his or her say at some suitable juncture in the narrative. But more important than this is the fact that we engage in a genuine dialogue with the speaker (and perhaps the other listeners). We comment as they speak ('Oh, *yes!*' ... 'They *didn't!*' ... 'Uh-huh?' ...), reinforcing what they say, encouraging them to continue. Like grooming, it's a way of saying, 'I'd rather be here with you than over there with Jim' – a statement of interest, a declaration of intent. But it is already a great deal more than any monkey or ape can do with grooming. For monkeys and apes, grooming is a strictly one-on-one activity. Indeed, actual grooming still is even with us: in most cases, those with whom we are intimate enough to engage in real grooming take it badly if we try to fondle someone else at the same time. The fact that grooming is a strictly one-on-one activity has very significant implications, given that establishing a relationship cannot be done simply by saying 'Let's be friends.' Rather, it requires an investment of time during which we literally *build* the relationship. With only so much time available for investing in that bonding process, this inevitably means there is a limit on the number of others that any one individual can hold in a relationship. In effect, if the same rules hold for human conversation as for primate grooming, then language allows us to treble that limit. Already we can see how language might allow us to extend the size of our social groups.

But language has other useful properties in this domain – not least the fact that, thanks to grammar, it can facilitate the exchange of information. One of the things that language allows us to do is to exchange information about the social networks into which we are embedded. We can catch up on news of Aunty Flo and Uncle Fred, find out where nephew Bill is these days, and why cousin Penny's marriage failed. Neither baboons nor chimpanzees can do that. What they do not themselves see with their own eyes they will never know about. If a chimpanzee's best friend reneges on the implicit coalition between them by buttering up to their mutual mortal enemy somewhere else in the forest, it will never know about it until the catastrophic day when the best friend supports the rival in a fight. But humans can find out. We can ask if anyone else has noticed anything untoward when we have even the slightest suspicion that something is going on; and others (anxious, perhaps, to curry favour with us) may whisper news of the dastardly deed in our ear before we are even aware of it – much as Iago sought to poison Othello's view of Desdemona in Shakespeare's famous play.

Language, in short, allows us to keep track of what is going on in the constantly changing world of our social relationships. Who's in and who's out, who isn't behaving as they should, who's showing signs of becoming a promising candidate to be our friend – or perhaps even a mate. That is hugely important for the effectiveness with which our relationships – and the entire social group in which we are necessarily embedded – works. It means that we have a better than fighting chance of working effectively within this social environment even when the size of the network is much larger than anything a chimpanzee could cope with. It means that, when we walk into a particular social event, we are primed with the most up-to-date knowledge of what everyone is up to. It is not totally foolproof, but we are less likely to make a naïve error than we would otherwise do.

A Uniqueness of Being

There was much excitement among psychologists during the 1950s and early 1960s at the prospect of being able to teach a human language to a chimpanzee. The issue hinged around the question of whether humans learn language instinctively or merely because growing up in a community of speakers offers the infant an unparalleled opportunity that it cannot avoid. Several families of American psychologists raised baby chimpanzees in their own homes, in some cases alongside their own offspring. Everything that was done to the human infant was done to the infant chimpanzee.

The experiments were in one sense an impressive success: the chimpanzees did learn to speak a few words of English. But in another sense, they were a dismal failure. The best the chimpanzee could do was to whisper sounds that resembled the English words it was imitating. And in the end, the much faster rate of development of chimpanzees meant that, rather than copying what the human child did, the chimpanzee simply acted as a very badly behaved model for the experimenter's offspring. Human children, it seems, are just superbly specialised imitating machines, and will readily learn to imitate whatever is on offer – especially if the model's behaviour is just a shade on the side of naughty. The experiments were abandoned and (with good reason) have never been repeated.

But there were perhaps two rather more interesting reasons why these experiments failed. One is that apes (and monkeys if it comes to that) simply lack the vocal apparatus to speak. The chimpanzee's larynx is set high in its throat just behind the back of the tongue, whereas the human's is set low in the throat (its top is marked by your 'Adam's apple'). Human babies are born with high larynxes, which drop as they grow into the language-learning period. This is perhaps just as well, because having a low larynx makes it impossible to swallow and breathe at the same time, as a result of which, adult humans are liable to drown themselves if they try to drink and speak at the same time. Because babies have a high larynx, they can breathe and swallow without drowning, which is convenient when you want to suckle hungrily at your mother's breast in the middle of the night when she's half asleep and not paying attention. Babies can keep going to the point of exhaustion, whereas an adult would have to keep breaking off to draw breath every minute or so. The whole process would be rather tiring, not to say frustrating.

The point of having a low larynx is that it greatly enlarges the resonance chamber in the throat and mouth, so enabling us to produce a range of sounds that is well beyond the capabilities of our ape cousins. Language without this enlarged articulation chamber would be an impoverished affair, to put it very mildly. In short, apes will never be able to speak because their vocal tract does not allow them to speak. Hindsight is a wonderful thing.

But there is another and perhaps more fundamental reason why apes cannot speak – and never will be able to. This is the phenomenon known as theory of mind that we encountered in Chapter 3. Ever since Noam Chomsky put linguistics on the map as a proper science half a century ago, linguists have focused their attention on grammar and how it allows us to transmit information by encoding what we want to say into the sound stream of speech. But difficult as the parsing of the grammatical structure of sentences may be, it is not perhaps the most difficult component of language. The real intellectual work of speech lies in our ability to anticipate how the listener is going to understand – or perhaps *not* understand! – what we have to say. If language was just about formulating correct grammatical sentences that described the world as we see it, then conversations would probably be the most boring thing we ever did. Just imagine it: 'Today, there is a new red stop sign at the crossroads ...' 'Well ... that's *very* interesting ... Hmm, I think I'll just have a word with Jemima over there ...'

What actually makes conversation interesting for us is that we play mind games. We tell jokes, we use metaphor so liberally that almost nothing we say has its literal meaning. 'When you leave, would you mind pulling the door behind you . . .' It would certainly raise some eyebrows if you did. And herein lies the big issue: we spend a great deal of effort during conversations trying to gauge just how our listeners will interpret what we say; or, conversely, trying to figure out just what it is that the speaker is trying to tell us. We speak in riddles and circumlocutions. We go to great efforts, it seems, to avoid saying *exactly* what we mean in plain simple English. For all the difference it would make, we might just as well be speaking French, or Polish or Chinese.

To be able to steer a way through this self-imposed confusion, we have to be able to put ourselves in the other person's mind and see the world from their point of view. In other words, we have to engage in deep mind-reading. Moreover, since our explanations of other people's behaviour naturally focuses on their motivations and intentions, we have to go beyond simple theory of mind (second-order intentionality) if we are going to be able to describe the behaviour of a third party. Iago had to *intend* [1] that Othello would *believe* [2] that Desdemona *wanted* [3] to love another before it was worth his while saying anything at all to the Moor. It was not what Desdemona *did* that was so important in the grand scheme of things, but what her actions implied about her *intentions*. It was Iago's instinctive ability to see how Othello would interpret the information about Desdemona's intentions – and the audience's ability to foresee the awful inevitability of that interpretation and anticipate its consequences – that makes the play work.

Had Iago not been able to engage in these mental gymnastics, he would not have been able to feed Othello a bunch of lies. Othello would have remained in sublime ignorance of Desdemona's presumed behaviour and he would never have killed her; nor, in his anguish at realising how he had misread the situation, would he have gone on to kill himself. In which case, the story would have lost most of its emotional force. Without third-order intentionality, Iago could not have done what he did. Without fourth-order intentionality, we the audience would not have been able to figure out the big story line. And without fifth-order, Shakespeare could not have put the whole thing together and manipulated our minds the way he so miraculously *intended* to do. Without fifth-order intentionality, he would have been the proverbial chimpanzee at the typewriter hitting keys at random and, most decidedly, *not* producing Shakespeare. Without theory of mind – indeed, without the higher orders of theory of mind – literature and much of everyday social intercourse would be impossible. We would live in a dull and intellectually impoverished world. Life would lose much of what makes it so interesting. And it probably would not be worth having a conversation with anyone.

In sum, the second reason why chimpanzees will never be able to speak is that they do not have the cognitive machinery to figure out the complexities of the mental world to the level that seems to be essential for fully fledged conversation between you and me about what some third party got up to in his or her relationship with someone else.

When did Speech Evolve?

If chimpanzees do not have the capacity for language and we do, when did language evolve? There are two approaches we can use to get at this question, though neither is entirely satisfactory on its own. The most obvious is to ask whether there are any anatomical correlates of language (or speech) that we might be able to identify in the fossil record. As it happens, there are, although they are somewhat indirect. The second is to exploit the relationships we found between neocortex size, group size and grooming time to ask when hominid group sizes would have been too large to sustain by grooming alone: that should be the point at which language had to evolve.

The first approach has been to examine some of the neural correlates of speech. One of these is the size of the hole through the bottom of the skull that the nerve to the tongue passes through. The size of this hole (the hypoglossal canal) reflects the size of the nerve, and the size of the nerve reflects the amount of work it has to do. Speech depends on precise articulation and this in turn depends on fine motor control of the tongue, jaw and lips in order to create exactly the right articulatory space in the mouth to produce particular sounds. Humans have a significantly larger hypoglossal canal than any of the African great apes (chimpanzees and gorillas). More important, all the fossil hominids after the appearance of archaic humans (the first members of our species *Homo sapiens* who appeared about 500,000 years ago) have hypoglossal canals that are similar in size to those of modern humans – and this includes both the Neanderthals and the Cro-Magnons (our own immediate ancestors in Europe). In contrast, all the australopithecine skulls in which this feature could be measured have ape-sized holes. The real problem is that there is a dearth of suitable skulls from which we can measure the size of the canal in between these two phases of our evolutionary history, so it is rather difficult to place an exact date on the point of the transition other than to say that it occurred some time between two million and 300,000 years ago.

A second study, carried out by Ann McLarnon at the Roehampton Institute, focused on breathing control. Modern humans, but not living monkeys or apes, have a dramatic enlargement of the vertebral canal in the region of the thoracic vertebrae in the upper chest. The nerves from this region control the chest muscles and the diaphragm, and are thus important in the fine control of breathing that is necessary to produce speech. Speaking requires us to release a steady, slow exhalation of air over a much longer period than is necessary for simply breathing. None of our primate cousins can do this, and they lack the enlarged thoracic nerves that would be required to control it. Examination of the thoracic vertebrae of fossil hominids suggests that this very conspicuous enlargement of the vertebral canal in this region does not appear until the same sort of time period as the enlargement of the hypoglossal canal. Older specimens, including both australopithecines and *Homo erectus*, all have thoracic vertebral canals that are, relatively speaking, no larger than those of other monkeys and apes. But Neanderthals and early modern humans from around 80,000 years ago all have canals that are indistinguishable in size from those of modern humans. Once again, however, we are left hanging uncertainly about the exact date because there are no fossil vertebrae from the intervening period. One thing that we can conclude, however, is that, since both Neanderthals and early modern humans both had enlarged thoracic vertebral canals, the most parsimonious conclusion is that they inherited this from their most recent common ancestor – archaic *Homo sapiens* who first appeared around 500,000 years ago.

Taken together, these analyses bracket the date at which speech evolved. The size of the thoracic nerve canal places the earliest possible date as some time after 1.6 million years ago (the last fossil in the sequence with an ape-like thoracic canal). Given that both Neanderthals and Cro-Magnons have modern-sized hypoglossal and thoracic nerve canals, the simplest explanation is that they inherited these traits from their common ancestor, archaic *Homo sapiens*. Hence, the latest possible date must be the appearance of that common ancestor, around half a million years ago.

An alternative approach to this problem is to see what we can learn from the relationship explored in Chapter 3 between neocortex size and group size and the fact that the amount of time spent grooming in Old World monkeys and apes is a function of social-group size. I discussed this at some considerable length in my book *Grooming, Gossip and the Evolution of Language.* The essence of my argument is that if we take the relationship between group size and neocortex size in primates and apply it to the fossil specimens, we can use it to predict how group sizes change across time for all the fossil hominids. Then, using these group sizes, we can exploit the relationship between group size and grooming time in Old World monkeys and apes to predict how much time each of these fossil populations would have had to spend grooming if it was to bond its groups in the conventional primate way. The results are shown in Figure 5.

What these analyses tell us is that required grooming time remains well within the limits for living monkeys and apes right throughout the australopithecine period of our evolutionary history. Only with the appearance of *Homo erectus* does it begin to rise above the figure of 20 per cent of total daytime that marks the upper limit for living non-human primates, and even then the rise is at first very slow. It is not until we get to the appearance of the earliest members of our own species (archaic *Homo sapiens*) 500,000 years ago that we find that the demand for grooming time has really taken off. It is only really at this point that the grooming time requirement seriously exceeds the limits we find in other monkeys and apes. The fact that this coincides rather nicely with the conclusions we drew from the anatomical evidence for speech reinforces the suggestion that language is a uniquely human trait.

In sum, then, it seems that speech (and hence language) must have been in place by the appearance of *Homo sapiens* half a

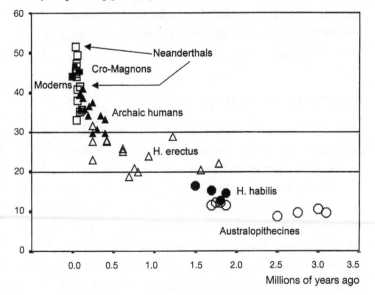

Figure 5: The amount of time that fossil populations of hominids would need to have spent grooming in order to bond their groups in the way that monkeys and apes do can be calculated from the known brain size of fossil specimens (shown in Figure 2, page 29), using the relationship between neocortex volume and group size, and the relationship between group size and grooming time for Old World monkeys and apes. The maximum amount of time that any monkey or ape species spends grooming is 20 per cent of total daytime. Humans would have to have evolved language by the time their grooming requirement reached 30 per cent (the upper limit at which conventional grooming plus chorusing would have allowed social bonding to be effective). (Source: Aiello and Dunbar, 1993.)

million years ago, at least in some form. Whether this would have been language as we know it today is a moot point. A plausible interpretation of the evidence suggests that speech/language did not evolve suddenly out of nowhere (as many linguists have assumed) but rather developed piecemeal to fill the bonding gap left by grooming once group size exceeded the size that could be bonded in the conventional primate manner. This raises the possibility that language actually went through a vocal phase that was not linguistic – in short, one that was musical rather than verbal. But before I turn to consider that possibility, there is one other curious feature of human conversation that I need to mention briefly.

Laughter: the Best Medicine?

Language has clearly been hugely successful in allowing us to get where we are. But at the same time, there is something missing from the story that I have sketched out here. And this has to do with the way grooming creates a bond between two monkeys. Being groomed seems to have an extraordinarily relaxing effect on our non-human cousins. During grooming, the heart rate slows and the animal visibly relaxes. Indeed, if it is groomed for long enough, the animal can actually drop off to sleep. The reason why grooming has this soporific effect is that it seems to be remarkably good at stimulating the brain to release endorphins, the brain's own painkillers. Endorphins belong to the family of chemicals known as opioids: they have a very similar chemical structure to the more conventional opiates like opium and morphine, which explains why we get addicted to the latter so easily.

Experimental studies of monkeys have confirmed that grooming triggers a release of endorphins. Moreover, animals that are given artificial opiates lose interest in grooming; and, when they are given opiate-blockers (chemicals like naloxone that lock onto the opioid receptor sites in the brain and prevent the body's natural opioids from producing their analgesic effect), they become increasingly restless and seek out grooming. Whatever else it does, grooming produces a sense of relaxedness and wellbeing in its recipients and it seems to be this effect that is the proximate mechanism that enables grooming to act as a bonding agent. We do not really understand how this works, but it is very clear that grooming acts as the immediate reinforcer that allows partners to feel good in each other's company. In some way, this sense of wellbeing is transmitted into a willingness to support each other in conflicts. We seem to act in much the same way: we are more willing to support or help out those whose company we enjoy.

This raises a puzzle. What in human interaction provides the chemical kick that does the same work, so allowing language to act as a bonding agent? Speech itself lacks the direct physical intervention to stimulate the opioid system in the way that grooming or massage does. Of course, we do resort to grooming – or at least what amounts to grooming – in our more intimate relationships. But that kind of mutual mauling is rather conspicuously restricted to our more intimate relationships – in fact, precisely the circumstances when we abandon language altogether. This equivalent of grooming (petting, stroking, fondling) is something we do only with our most intimate associates: it is pretty much confined to mates, parents and children, less often to grandparents and one's very best friends, much less often still to more distant relatives like aunts, uncles, nieces or nephews and cousins and almost never to anyone else (except other people's babies). Such attention directed to one's doctor, teacher, pupil or work colleague, or worse still a stranger on the street, is likely to raise eyebrows at best, and produce a deep sense of outrage in the recipient at worst – and, these days, probably a court summons for harassment.

This brings up an interesting question as to why we find intimate contact between strangers so disconcerting. My guess is that it has to do precisely with the fact that we find close physical contact deeply arousing and that, for us humans as with the bonobos, there is only a very fine line between physical contact of this kind and fullblown sex: the one spills all too easily over into the other precisely because physical contact in a relaxed situation is emotionally arousing. So how do we create bonds of intimacy with those with whom we do not wish to have sex at the drop of hat? The answer, I am going to suggest, is that we make them laugh.

Laughter, when you think about it, is a remarkably odd behaviour. Chimpanzees have something vaguely like it, and this is indeed thought to be the origins of the human laugh. But it is pretty much restricted to play situations. When inviting or engaged in play, chimpanzees often give a series of quiet rapid pants with an open mouth expression (termed the 'relaxed open mouth' or ROM face). But this behaviour resembles only the most basic forms of human laughter, more the kind of genteel laughter that tinkles among the teacups in polite society or the kind that young children use when inviting play. Humans also engage in much more forceful kinds of laughter (as in 'belly laughs'), and they do it in a far greater range of circumstances than do chimpanzees (or, for that matter, young children). No one else in the animal kingdom laughs (at least, not in the extensive way that we do).

So what is going on here?

Well, think about what happens when you laugh, and particularly when you laugh heartily, letting go all your inhibitions and having a good old roar. You come out of it feeling . . . well, a little light-headed, certainly much more relaxed, and generally rather at peace with the world. Sound familiar? Well, of course it is: it is the endorphin story all over again. Laughter seems to be a good releaser of endorphins. And there is some indirect experimental evidence to support this. The evidence is indirect because it is difficult to measure endorphin production directly (it requires the rather unpleasant procedure known as a lumbar puncture, in which a rather large needle is shoved forcibly into the space between two adjacent vertebrae). Most studies have therefore used pain tolerance as a more easily assayed measure. The logic is that, if endorphins are a part of the pain-control system, then the more endorphins released the more pain you will be able to put up with.

My students Julie Stow and Giselle Partridge carried out two separate experiments to try to test this idea. In these experiments, we asked subjects to see how long they could keep a frozen wine-cooler sleeve on their arm. Once they had done this, they were shown a video clip from either a documentary or a comedy programme, after which they were asked to try the wine cooler again. Subjects who were shown a boring documentary showed no increase in how long they could stand the pain afterwards compared to what they had managed before. But those who watched a comedy were able to stand the frozen wine cooler for significantly longer than they had previously done. Moreover, the increase in tolerance that they showed was related to how much they had laughed during the video: those who laughed more were more tolerant of the pain than those who had laughed less.

Perhaps this explains another odd feature of our conversational behaviour – the fact that we seem to spend a great deal of our time trying to make each other laugh. It seems rather odd that we have a mechanism (language) specifically designed to allow us to exchange information, yet we seldom seem to use it for such a solemn purpose. Indeed, in all but the most exceptional circumstances, we find it rather boring if those we engage in conversation insist on producing an unending stream of worthily stolid information. 'About that new red stop sign I noticed down at the crossroads . . .' 'Uh-huh? . . . Now, which way did you say the bar was?' But, start talking to someone who feeds you one-liners or who peppers their conversation with witticisms and, all of a sudden, the bar seems to lose its magical appeal.

And this is exactly what we found in a study of conversations carried out by Feroud Seepersand. He listened in on natural conversations in bars and cafés, every 30 seconds making a note of the topic being discussed, while at the same time keeping a record of when the individuals laughed. His results showed that a pair continued talking about the same topic for significantly longer after one of them had laughed than if neither had laughed. Like grooming, it seems that laughter encourages you to stay put and continue the interaction with a particular partner. It floods the brain with endorphins and just makes you feel positively disposed towards the other person.

In fact, there is some recent evidence that puts all this into an even more interesting light. A study of patients with damage to different parts of their brains has revealed that a particular area in the right frontal lobe is crucial for the appreciation of humour. You can be missing almost any other bit of the brain (including bits on the left side) and yet still appreciate humour. More extraordinary still is the fact that this includes not just cartoons and other 'visual' humour but also verbal humour – the very thing we might suppose would be handled by the language areas in the brain's left hemisphere where our speech centres are located. People with this same crucial bit of the right hemisphere missing also show greatly reduced laughter and smiling responses. Interestingly, this part of the right brain also has direct neural links down to the amygdala in the limbic system, the part of the brain that is especially involved in processing emotions and emotional cues.

Laughter is a ritualised activity that is highly contagious. We rarely laugh when on our own: indeed, those who do so invariably attract adverse comment. As convention has it, only the mad laugh on their own; the rest of us laugh because others laugh or because social situations are particularly prone to triggering laughter. That's why canned laughter on TV gets us going when without it, sitting alone in our room watching TV late at night, laughing is probably the last thing we would bother to do. It is also why, when someone tells a joke in a foreign language, we will laugh heartily with everyone else despite the fact that we have not understood a word. It is this chorusing feature of laughter that attracts my attention here, not so much because of the laughter but because of the chorusing that it seems to involve.

It seems that, at some point during the course of human evolution, we borrowed the chimpanzee playface and its associated vocalisations and exaggerated them to provide the reinforcer for grooming at a distance. Since the brain areas involved in laughter and language seem to be very different – indeed, they are not even in the same hemisphere of the brain – laughter may well have evolved long before language. The fact that laughter is so contagious perhaps suggests that it was used in a kind of communal ritual alongside non-verbal vocalisations like conventional primate contact calls. Later, of course, the acquisition of language allowed us to use verbal constructions to stimulate laughter in others more effectively. Jokes, it seems, have a very ancient heritage, much older in all likelihood than anything else we do with language.

Trip the Light Fantastic

There seems to be something very fundamental (in a literal sense) about music and song. We rise to it emotionally in a way we seldom rise to mere words. Composers since time immemorial have recognised that they can stir our emotions by the way they order the sequences of toned sounds, producing now a sense of joy, now despair, or exciting us with the rhythms that set the feet a-tapping. There have inevitably been arguments as to whether this emotional manipulation is culture-specific or not. Do up-turns in tone make all humans joyful and downturns make us all despair? Do major keys raise our hopes and minor keys dash them? Does fast music make us feel excited and roused, and slow music make us feel languorous? Or are these simply associations we have learned from the last dozen or so centuries of western music?

I am less interested in the answer to this question than in the fact that composers can manipulate our emotions at all, irrespective of the origins of the particular code they may use to do it. It seems to be a remarkable universal of human nature that we respond emotionally to music in this way, and that we are especially prone to do so when we do it in groups. Communal singing, as almost all religions have long recognised, seems to have a particularly strong emotional hold on us.

Why should music do this to us and what role has it played in the story of human evolution?

The answer to the first question is still shrouded in mystery. But it seems that some musical tones do trigger off deep responses somewhere in the brain. Aside from the more obvious activity to be expected in the auditory cortex where all sounds are processed, the main responses are in the right hemisphere and in regions in the evolutionarily more ancient limbic system.* Since that is the opposite side of the brain to where language has its main centres (the left hemisphere), it seems plausible to infer that music and language have had separate evolutionary histories. Indeed, the deeply emotional stirrings generated by music suggest to me that music has very ancient origins, long predating the evolution of language, and this perhaps gives us a clue as to how we might answer the second question on music's role in our evolutionary history.

The answer, I think, lies in the fact that something similar to

---

* One recent suggestion has been that our sensitivity to music derives from the vocal exchanges that occur between all mammalian mothers and infants. If so, then our responsiveness to music has very ancient roots indeed.

non-human primate contact calling must have bridged the gap between the first rise in group size above the conventional non-human primate limit (about 60–70 individuals) and the rise of true language (once group size had exceeded around 120). Given what we know both about primate contact calls and their use in choruses and about music, it seems increasingly likely to me that it was singing that filled this gap.

Singing is a form of vocal activity that lends itself to multi-tasking and the double use of time. We still do it. From the unique women's *waulking* songs of the Outer Hebrides to sea-men's shanties, and from the marching songs of armies to foot-ball fans singing on soccer terraces, singing rouses emotions and binds members of the group while they are engaged in some other activity that prevents more intimate forms of con-tact. Of course, singing also helps while away the time and makes a hard task more bearable. But ask yourself: how does it achieve this? It is surely not just by keeping the mind busy while the hands haul on the ropes! My guess is that it's because com-munal singing triggers the release of endorphins and it is these that make the work seem lighter.

That endorphins might in fact be involved has been known for some time. In one experiment, subjects listened to tapes of music, and indicated when they felt a thrill of excitement at a particular passage. The pattern of thrills was quite consistent from one day to the next for any individual subject, even though, as we might have expected, people varied enormously in terms of which particular passages triggered these thrill episodes. However, if subjects were given an injection of nalox-one (the same antidote to endorphins that blocks monkeys' sat-isfaction responses to grooming) between successive auditions, they failed to show such marked thrills on the following audi-tion as they had on the previous control exposure. Those who had an injection that contained nothing but saline fluid exhibited no differences between successive auditions. This is strong circumstantial evidence that endorphins are involved.

Why and how singing has this effect is as yet a complete mystery. Very little work has been done in this area to date. Nonetheless, the prima facie case for this hypothesis is very strong. And it *feels* right. Of particular significance here is the fact that we can induce these emotional effects by music alone without the need for any words. Wordless songs and the pure tonalities of musical instruments produce the same effects as the most rousing lyrics. Gregorian plainsong of the Catholic monastic tradition provides a particularly obvious example of this. It is the sounds of harmonious chant that we find so compelling, not the words – especially given the fact that most of it is in ancient Latin and not understood. In fact, so important is the sound and so unimportant are the words themselves that during the early polyphonic period of European music (around the twelfth to thirteenth centuries) composers often did not bother to pay too much attention to how they used the lyrics provided by poems or a text from the Bible. It was by no means uncommon for the lyric to end mid-sentence – sometimes, even halfway through a word! – if that suited the music better.

These observations allow us to make sense of the kinds of phenomena we see in contexts like Pentecostal church services. Here, the musicians, choir and minister create an increasingly intense and rousing musical torrent that gradually draws the congregation, one by one, into the flow of the excitement until everyone is waving their arms, jigging their bodies, and bursting into 'Amens!' and 'Hallelujahs!' at appropriate moments. Some even appear to get carried away into trance-like states. So compelling is the music, that it is difficult even for sceptical non-believers to resist joining in – just as it is difficult to sit still while listening to an Irish jug band playing reels and jigs in a pub.

My guess is that very early on, song became wrapped up with dance. We seem to respond with especial enthusiasm to the rhythmicity of dance, and dance is of course widely used in the rituals of both traditional societies (think of the trance dances of the !Kung San bushmen) and advanced religions (think of the way the priests danced before the Ark of the Covenant in the Judaism of King David's time, and still do, more than two and a half millennia later, in the swaying dance of the *dabtara* or deacons in the Ethiopian Coptic Church). Indeed, dance has been exploited very specifically to induce states of euphoria and trance among the 'whirling dervish' sect of Sufi Islam: in this case, the dancers spin round in unison – an impression that is exaggerated by the long white over-garments they wear – until they collapse into a trance-like state.

Are these trance states some kind of self-induced opioid high? Is this why we so enjoy dancing, a phenomenon that probably ranks, along with smiling and laughter, as one of the most futile of all human universals? Were dance and singing, and perhaps the rhythmic clapping of hands that so often accompanies both of these, an early supplement to physical grooming that allowed *Homo erectus* to enlarge its groups beyond the limit imposed by the immediate time constraints on grooming?

Music-making using instruments was presumably a much later invention, one that occurred long after the rise of singing and perhaps even tens of thousands of years after the appearance of speech and true language. The vast array of musical instruments with which we are now familiar is, of course, of very recent origin. Examples of stringed and brass instruments, as well as drums, do not appear in the archaeological record much before a few thousand years BC. However, simple wind instruments of the flute or recorder type have a much more ancient history. One beautiful example carved from deer bone was found in the accumulated debris of a Cro-Magnon occupation site in a cave floor in France dating from some 30–40,000 years ago. As it survives, this instrument has four holes on the front and two on the reverse, and was clearly designed to play a pentatonic scale (which it still does quite admirably). Another flute, made out of cave bear bone, was found at a Neanderthal site in modern Slovenia dated to 53,000 years ago. Reconstructions made from original materials (real cave bear bones) play well and a competent flautist can coax out of these instruments almost the full range of sounds that can be produced from a modern recorder. Making these instruments using contemporary tools is a laborious business, so our prehistoric ancestors must have viewed the effort especially worthwhile.

We clearly differ from our ape and monkey cousins in our use of language. However, many of the core features of language, and the associated non-verbal components that make conversation possible, bear important similarities to the kinds of social communication we find in other primates. That we use language to exchange complex technical information is undoubtedly important, but it seems likely that this was a relatively recent development. Speech and language evolved to enable us to bond social groups that were getting too large to bond by conventional primate social grooming. We seem still to use it mainly for these purposes. Moreover, in order to enable language to do this job effectively, we have to draw heavily on some non-verbal features (laughter and music) that take us straight back into the chemical processes that underpin grooming. However, with laughter and music, we are at last beginning to find elements which, if not uniquely human, do at least find expression among humans with a frequency and intensity that are perhaps unique.

Language and music raise for us another important feature

of human nature, namely the whole complex business of culture. If culture can be said to be the hallmark of humanity, then language might be said to be its handmaiden. But what is this thing called culture? And are we the only species that can lay claim to it?

6 High Culture

His reverie broken, the artist looks around the clearing. The man by the edge still works at his stave, shaping the new spear onto which he will later haft a flint spearhead; the two women work at their skin, silently focused on a task that requires both physical effort and concentration. He walks across to where a fire burns between some stones. Above it, the remains of a deer haunch smoke in the heat of the fire's embers on a wooden spindle. He had not realised how long he had been at the cave. It is already late afternoon, and he has not eaten since the morning. He pulls a loose strip from the haunch and eats, savouring the meat.

As he crouches beside the fire chewing on the deer meat, he notices the two girls coming back down the path into the camp, laughing and giggling. They return to where they had left their pretend doll asleep. One bends to pick it up, cradling it in her arms as she would a baby. That maternal action causes him to remember something he had quite forgotten about. He had been whittling away at a piece of antler yesterday for them. He gets up and goes to the crude skin-covered shelter on the edge of the clearing that he and his family share at night.

Ducking down, he enters. In the half light inside, he rummages about on the far side where the skin covering has been pegged to the ground beside one of the willow uprights that forms the frame.

It takes him a moment or two to find what he is looking for under the fur cape that he uses when he is out hunting in cold weather. Then his hand grasps the cold haft. He withdraws a piece of deer antler about nine inches long that has been carved to resemble a body. One end, where the stem widens into a broken tine, has been deftly remodelled to resemble a head, with tiny nose and mouth. Incisions down the length of the stem etch out arms and legs. He has a little more work to do to finish it off, and then he will have a fine doll for the girls.

The idea came to him a few days ago as he sat watching the girls playing with their pieces of wood. He marvelled then at their ability to imagine that a simple unfashioned twig could be a baby to be cared for and cosseted as one might the real thing. Later, when he was searching through a pile of spear ends and other bric-a-brac, he came across the piece of antler, cast aside from some other task. And in it, he had seen in his mind's eye for one fleeting moment a baby's head. He had smiled, thinking how much pleasure the girls would get if he could fashion the antler a little to make it more baby-like, giving it more of a nose and sketching in its arms and legs.

But he did not quite finish it yesterday, so he had hidden it in the shelter awaiting an opportunity to finish it off. Now he rummages under the bearskin for the flint blade that he uses for cutting carcasses. Taking the blade and the doll-to-be, he leaves the shelter and walks down the path that leads back up into the valley. A few hundred yards from the camp, he settles down on a rock just above the riverbank and sets to work easing slivers off the antler with the flint blade. In an hour he is done to his satisfaction. The doll now has a finished head with a tiny hooked nose and just a hint of hair etched out with criss-crossed incisions around the back and top of the head, arms and legs incised into the stem and – he is especially pleased with himself for this final stroke of genius – a tiny belly button.

He gets up, brushing the scraps of antler off his lap as he does so, and walks back into the camp. The girls are still engrossed in tending to the moss crib. He calls them over, holding out the doll for them. They come running, not quite sure what he is showing them.

'I have made you a baby,' he says, holding the doll up so that they can see the head and body more clearly.

The girls gasp with delight and run to take it from him. The older girl cradles it carefully in her arms, cooing over its head while the younger reaches out to touch it gently on the face.

Culture is everywhere associated with human societies. There is no tribe or nation-state on earth that does not lay claim to having a culture of its own. Anthropologists have always seen culture as their particular turf, the very core of their discipline, so they would seem the obvious people to turn to in order to find out whether or not it is a uniquely human characteristic. Unfortunately, it seems that anthropologists cannot agree among themselves on how to define culture. In a celebrated paper published more than half a century ago, two very eminent American cultural anthropologists, Alfred Kroeber and Clyde Kluckholm, surveyed the literature and concluded that their anthropologists had used the term 'culture' or 'cultural' in at least 160 different ways. Not very helpful.

If we sift through these definitions a bit more carefully, however, it seems that we can reduce the confusion to three broad but fairly commonsensical themes. One set of definitions defines culture in terms of the ideas in people's minds: these include not only instructions on how to do things, but also statements about the meaning and significance of life, or things like rules for the behaviour of different classes of people within a society and the rituals of public religion. This is the sense that social anthropologists most often use when they talk about the world of tribal societies. It concerns the rules and principles that keep the society together as a cohesive unit because its members share a common view of the world. The second cluster defines culture as material objects. This is the sense that archaeologists most often use when they examine the buried remains of past civilisations. Culture in this sense is very solid – the pots and pans, the jewellery and weapons, the house or tent designs, the dolls and statuettes, and all the other inessential paraphernalia of everyday life . . . the bric-a-brac that we leave behind us when we die. The third cluster of definitions is of 'high' culture, culture in the everyday sense – the arts, music, literature, learned journals. A great deal (but perhaps not all) of this is dependent on language as a medium for both expression and transmission. We can teach children how to make pots and axes merely by showing them how to do it; but we cannot show them how to write novels unless they first have language and understand what it means to express thoughts in words.

The Language of Culture

Anthropologists have always been firmly resistant to the idea that anyone other than humans has culture. Culture, in their view, is *the* uniquely human trait, that which marks us out as a cut above the dumb beasts, the one thing that allows us to distance ourselves from our biological roots. So far as they are concerned, the beasts are locked into the cage of their biological dispositions, driven to behave in some mindless way by the dictates of their genes. But humans are different: we can rise above our genes, and act in ways that are at odds with the demands of our biological heritage. We can voluntarily join monasteries and decline to reproduce, contrary to the dictates of our biological drive; we can commit suicide at the behest of cultural perceptions of honour or religion. And all this is

141

because we alone have Culture with a very capital C. Although anthropologists have not often exhibited much interest in the brute beasts (other than as chattels of the peoples they study), on the rare occasions when they have done so their view has invariably been dismissive. They have usually been content to rest their claim that only humans have culture on the view that language is essential to culture and only humans have language ... QED.

If language does underpin culture, then its absence in animals would oblige us to rule out the possibility that animals have culture at least in the third (literary) sense. Even chimpanzees do not tell stories of the old times round the camp fire, entertaining each other with tales of ancient heroes and frightening children with stories of malevolent spirits lurking in the trees just beyond the circle of firelight. So the great apes fail the third test for culture. But they do so on the slightly contrived grounds that they cannot speak.

And that raises the key question: just how essential is speech (or, alternatively, language) for culture? I might be happy to agree that literature and art are unique to humans, but is this only because humans have language and other species do not? What about the other definitions? One could legitimately argue, for example, that language is relevant to the *transmission* of culture in humans, but not to its actual *creation*. The creation of an artefact or a rule or a poem depends on something deeper than the mere mechanism to pass it on to others – despite the fact that, as everyone would surely agree, culture simply isn't culture if it is not shared with others. In other words, there are still at least two other senses of culture, and great apes might yet pass either of these tests. What can we make of material culture and the world of ideas?

Tools and material culture have been the bedrock of the fossil-hunters as well as archaeologists since we first started to take an interest in our past. Searching through the detritus of ancient campsites, they have always emphasised the significance of stone and bone carved into purposeful shapes. Yet it is not until the appearance of *Homo habilis* around two million years ago that we find undisputed evidence for manufactured tools (although it has to be said that the provenance of even these has been doubted of late). The australopithecine fossil record is almost silent. To be sure, there are broken stones that have been interpreted as tools, but informed opinion favours the view that these are, if anything, stones used as tools rather than stones deliberately *manufactured* into tools. And the emphasis here is very much on the word 'manufactured' – it is one thing to pick up a stone and use it as a hammer to crack nuts (to solve a technological problem of the moment), but it is entirely another to imagine within its rounded shape the form of a Venus figurine and then chip away at the stone until that form is revealed.

However, we need to be careful not to get distracted by the fossil record here. When archaeologists discuss ancient tools, they focus almost to exclusion on stones and bones, items that preserve well in the fossil record. What about tools made of wood or other plant material? The oldest recorded spear is a length of wood whose tip had been hardened in a fire that was recovered from gravel beds near Clacton in southern England and dated to around 400,000 years ago. That is more than one and a half million years after the first certainly manufactured stone tools appear in the fossil record. Might sticks, long since dissolved into their constituent atoms, have been used as tools even before stones were chipped carefully into predetermined shapes by more advanced hominids? Those who defend the claim that great apes do not have culture have focused almost to exclusion on stone tools. Have they overlooked a crucial part of the evidence? The next section explores evidence of tools used by the great apes – but first, what of the third suggestion, that culture is about the ideas in people's heads?

This is perhaps the trickiest of the three definitions, because we seem, at least for the moment, to be barred from knowing what goes on in another animal's mind. Indeed, strictly speaking, we are even barred from knowing what is going on in another human's mind. If it is sometimes difficult for us to be sure of our own inner experiences, we are at an even greater disadvantage when it comes to knowing how another individual feels or thinks. How can *I* know what someone else's cultural belief means to *them*? Most often, anthropologists just ask people about their views and do not worry too much about the philosophical niceties. 'So what *do you* believe about the origins of your people? What *do you* believe about life after death?'

That is fine for humans, since we can ask each other what we think and figure out at least some (even if only indirect) evidence as to whether any of these assertions are true. We do that mostly by a kind of argument by analogy. I go 'Ouch!' and screw up my face when I hit my thumb with a hammer, and assert all sorts of things about what I feel; if you behave in the same way and say the same kinds of things when you hit your thumb, I conclude that you must be having the same kinds of feelings that I do. So if you tell me that you believe that demons live in the spring, I can believe you because I have some sense of what you might mean by this claim, and enough evidence from more workaday events to assume that our minds work in similar ways. However, this argument by analogy does not really help us when it comes to species that do not have language – our occasional assertions about our pets notwithstanding. We really *cannot* get inside their heads, no matter what we think.

But more important, perhaps, we come back to the point that language is simply the means by which we pass on cultural behaviour and instructions. It is not culture itself, and it does not determine culture as such, though the kinds of words we have may constrain or limit what knowledge we can pass on. It would seem invidious to assert that animals lack culture merely because they lack language, as though lacking the ability to write might in some sense deny one the capacity to imagine stories. The cultural bit is, surely, contained in that leap of imagination in which the fictional event is conceived and elaborated, not in the fact of consigning it to paper. The storyteller's art lies in the creation of the story, not in its being told per se. Of course, the telling makes it social, thereby making it a part of our cultural heritage. But, there is more to the story than simply telling it.

So we need something more concrete and observable that we can apply to species that do not have language. One solution has been to focus on the rituals of life – the meat and drink of anthropologists who study the exotic customs and unexpected habits of tribal peoples in far-flung parts of the world. Here, we are perhaps on slightly firmer ground because rituals and behaviour are things that we *can* see. The difficulty, of course, will lie in the kinds of inferences we can draw as to exactly what lies behind the behaviour. And it is here that we come up against the psychologists. Their main concern is not with what culture is but with how it is transmitted from one individual to another. Their view is that culture is something that is learned by a particular social process, and they are not too bothered about what kinds of things might be involved.

But before we leap into that particular pond, let us see what more we can learn about ape material culture.

Of Sticks and Stones

Bill McGrew has the distinction of having a foot in both camps. He is an ethologist whose original PhD was on the playground behaviour of young children, but he also has a PhD in social anthropology based on a study of the cultural behaviour of chimpanzees. His main concern has been with what he sees as the generic chauvinism of the anthropological fraternity – their claim that *only* humans have culture. His aim is to show that the way anthropologists use the term culture in the case of humans would not in fact allow us to distinguish between the *material* culture (the stones and bones side of the story) of humans and that of the great apes. The burden of his argument is that the kinds of tools that chimpanzees use in their everyday lives (twigs used to fish for termites, sticks used as hammers) are just the kinds of thing that do not preserve well. Yet many of the tools and artefacts made by modern, never mind long-extinct, humans are of just this kind. That we do not see them in the archaeological record does not mean that they were not once there.

So what kinds of things are we talking about? What *is* chimpanzee culture in this sense?

We can perhaps distinguish two different kinds of things that might be of interest. One is objects that are used – sometimes manufactured – to enable the animals to solve a problem (in the chimpanzees' case, usually something to do with food). The other is rather more nebulous, and has to do with habits or behaviours that are present in one population of apes but not in a neighbouring one.

Of the first of these forms of culture in chimpanzees, their tool-using and -making activities, perhaps the most famous involves termite-fishing. Jane Goodall's original observations of termite-fishing at Gombe in the mid-1960s were little short of sensational. Termites are considered a delicacy by humans and animals alike, but their concrete-like nests make the insects all but impregnable. Only when they swarm in their annual mating flights at the start of the rainy season each year are they vulnerable to predators – human and animal – who grab them as they fly or pick them up as, later, they struggle on the ground after they have lost their wings. At Gombe, however, Jane Goodall discovered that her chimpanzees had solved the problem of how to get termites at any time of year. They did this by selecting long grass stems (occasionally fine twigs), stripping them of their side branches and leaves, and then carefully poking the stems down an entrance hole in the termite nest. The soldier termites immediately rush to attack the intruder, grip onto the stem with their pincer jaws and hang on grimly as the chimpanzee carefully withdraws the stem. It is just a matter of pulling the stem through the teeth to get a succulent protein-rich mouthful.

Later, Goodall witnessed them fishing for safari ants in a similar way. Safari ants are, as any one who has spent time in Africa will know, creatures to be treated with the deepest respect. From time to time, these ants set off in massive columns containing as many as a million individuals in search of a new nest. Anything that falls in their path is instantly bitten to death by the ferocious centimetre-long soldiers with their daunting pincer jaws, and then devoured by the workers and reproductives that follow on behind. Even an antelope carcass can be stripped of flesh in a matter of hours. The Gombe chimps had discovered that, while standing a respectful distance clear of the column of ants, they could lean over and poke a stick into the middle of the column; then, as the soldiers raced up the stick in search of the offending source of the disturbance, they would withdraw the stick, run the finger and thumb of the other hand quickly up it to gather the ants, shove the bundle into their mouths in one fluid movement and chew like fury. Result: a tasty nutritious meal, though one that has a certain frisson of excitement afforded by the prospect that any lapse of skill will result in one being bitten in the most delicate places.

These particular activities turned out to be unique to the Gombe population. Neither occurred in the Mahale population of chimpanzees just a hundred miles further south on Lake Tanganyika. However, similar behaviours did occur in some West African chimpanzee populations, but with important differences. Rather than using long, thin grass stems to fish for termites, West African populations might chew the ends of sticks to make a kind of toothbrush which the termites could grip on to. These subtle differences emphasise the key point that these are cultural habits confined to particular populations. At some time in the past, one chimpanzee invented a particular technique, and that technique was copied by those around it. The habit gradually spread through the community, but its form remained relatively stable over time because the animals simply copied what others do and did not add to it by their own trial-and-error experiences. Long after the technique's original discoverer died, its descendants still fish for termites or break open nuts in pretty much the same way it invented. The technique appears to have been handed down from one generation to the next by the simple process of copying. Even though the original inventor might have arrived at the technique by a process of trial and error, later generations simply copied what they saw.

The fact that techniques are identical within a population but differ between populations is evidence that these are genuine socially-learned habits and not something that each animal has learned for itself by trial and error. If trial-and-error learning were involved, we would not expect to see differences just between populations: instead, we would see several different techniques being used to solve the same problem within each population – the product of individuals' own unaided efforts at trying to find a sensible solution to a tricky problem (and hence more or less the same range of behaviours across different populations.

Sceptics might, however, point out that, although tool using (and even tool making) turns out to be common within chimpanzee communities (and perhaps even among orang-utans), the list of tools is not that impressive – a mere dozen or so sticks, stones and leaves. Leaves, for example, have been observed being used to wipe away blood and mucus from wounds or as sponges to soak up water to drink from an otherwise inaccessible tree hole. That's just the kind of use to which humans might put leaves. But should we be impressed? If we compare the range of artefacts that chimpanzees exhibit with the range that modern humans have, we might legitimately be scornful of any similarities. Where are the ploughs, the donkey carts, the bows and arrows – never mind the hanging gardens of Babylon or the Great Pyramid of Giza?

McGrew would, however, caution us against being too hasty. If we compare the chimpanzee's toolkit of a dozen or so documented items against the toolkits of the least technologically advanced human cultures, they do not come out quite so badly. Before they were finally exterminated by the white colonists towards the end of the nineteenth century, the native Tasmanians lived a very basic – not to say primitive – life as hunter-gatherers. They had been isolated from the rest of Australian aboriginal culture for around 10,000 years by the sea-level rise that had occurred following the melting of the polar ice caps at the end of the last Ice Age. Cut off from cultural diffusion, they seem to have lost the habit (or even need) for many of the basic artefacts common among mainland Australian aboriginal populations, while those invented later just never made it across the Bass Strait. They lacked such items as pottery, ironware, bows and arrows, fish-hooks, spear-throwers, boomerangs, and canoes, all of which were common on the mainland (and many of which they had in fact themselves possessed during the early stages of their own history, according to the archaeological record). So far as we can ascertain, their entire tool-kit at the end of their history consisted of a mere eighteen items – digging sticks, some very basic stone implements, spears, grass rope, baskets, hides (from which to ambush prey) and traps (for birds). In short, a list that is not so different in size or content from the accredited list of tools for modern chimpanzees.

If we make allowances for qualitative differences in the toolkits of chimpanzees and those produced by modern humans, there are really only two things in the Tasmanians' toolkit that chimpanzees do not have – containers for carrying things (such as baskets or gourds) and structures (things like hides and traps). For the rest, says McGrew, if you were to see a chimpanzee tool in a museum display case without its label, you could never know for sure whether it came from a human or a chimpanzee population.

The same might be said of the stones that the chimpanzees of West Africa use as hammers to crack open the rock-hard nuts of the Guinea oil palm. Because the chimpanzees remember where they last used the stones and collect them on the way to exploit another palm tree, the same rocks are used over and over again. As a result, the hammer rocks acquire distinctive wear patterns that have some slight resemblance to the wear patterns of the very earliest stone tools used by our hominid ancestors around two million years ago. So on shape and wear patterns alone, we might question whether we can be absolutely sure that the ones in the fossil record were really created or used by hominids and not by apes. Archaeologists have always assumed they are the product of hominid activity, marking in some sense the appearance of a particular kind of more advanced humanlike mind. But are they right to make that assumption? The answer, of course, is that we cannot be absolutely sure, and maybe we should just be more cautious about what we infer from the archaeological record.

More recently, a careful review of chimpanzee culture has listed more than 39 tools and behaviours that occur in some but not all populations of chimpanzees, and are thus putative candidates for being cultural phenomena. Besides the sticks and stones with which we are now familiar, there are little quirks of behaviour that are unique to individual communities. One of these is a curious form of mutual grooming found in only four of the seven sites surveyed. Here, two chimpanzees sit facing each other and lean the wrist of one arm against their partner's wrist and then each proceeds to groom its companion's raised forearm. This innocuous behaviour is of interest precisely because it seems to be a habit peculiar only to some chimpanzee communities (such as the Mahale and Kibale communities in East Africa) while being completely unknown from other nearby communities (such as the Gombe and Budongo sites). It is the chimpanzee equivalent of the differences in the way humans point to themselves: some (such as most Europeans) point at their chest, some (like the Japanese) at their noses. Or the equally surprising differences among Europeans in the way they beckon: northern Europeans do so palm up, but southern Europeans and Africans do so palm down. Since we would have no hesitation in counting these variations of behaviour as cultural differences between human societies, McGrew argues that we should do the same for chimpanzees. What is good for the goose, as it were, must also be good for the gander – otherwise we are guilty of the unforgivable sin of moving the goalposts in an attempt to preserve our uniqueness as humans.

The Doctor's Round

One phenomenon that has attracted a great deal of attention has been the use of medicinal herbs by chimpanzees. Studies carried out in a number of chimpanzee communities have revealed that the animals make use of a surprising range of natural medicines that have genuinely beneficial effects on their health. Indeed, many of these same herbs are used by local tribal peoples for the very same conditions for which the chimpanzees seem to use them. Michael Huffman and his colleagues from Japan's Primate Research Institute have documented some 28 species of plants that chimpanzees eat apparently for their medicinal rather than their nutritional properties. These include berries, leaves, pith and bark that are often bitter to taste or, in the case of leaves, hairy and awkward to swallow; in many cases, the animal chews or sucks on the plant part, swallows the juice and spits out the fibrous residue. Some of these plants are known to contain active ingredients that are poisonous to a variety of parasites (including helminths, a class of intestinal worms that commonly trouble both human and ape populations in Africa). In Ghana, for example, the bark of the species *Entada abyssinica* (a member of the mimosa family) is used by the local humans as an emetic and a cure for diarrhoea, and the same plant is eaten by the chimpanzees at Mahale in East Africa for what seems to be the very same condition. The bark of this species has also been shown to have significant plasmocidal and antischistosomal properties. (Schistosomes are another major family of parasites that includes the liver-fluke and the agent that causes the debilitating snail-borne disease known as bilharzia, both of which are major diseases of rural African populations.)

Chewing the bitter pith of the *Vernonia* plant has been observed among both the Mahale and Gombe chimpanzee populations in Tanzania and by those inhabiting the Kahuzi-Biega region of the Congo. Although widespread throughout sub-Saharan Africa, plants of this family are relatively scarce locally, and often require detours from the animals' ranging path to locate them. Moreover, when consuming the pith of these plants, the animal first carefully removes the outer bark and leaves to locate the inner pith. That this is deliberate behaviour is suggested by the fact that other animals accompanying the eater usually ignore what it is doing, and feed instead on more conventional food plants. The animals that target these bitter plants tend to be those already showing signs of intestinal disorders or parasite infestation. After they have eaten these plants, their health invariably shows a marked improvement.

What are we to make of this behaviour? If the use of medicines is a part of human culture (and there are obviously very good reasons for saying that it is), then can we in all honesty deny the same claim for our non-human cousins?

Some would want to argue against this suggestion, mainly on the grounds that if we are to credit animals with culture in the human sense, we must be absolutely clear that we are talking about the same kind of thing that we mean in humans. One problem identified here is that it is often hard to distinguish between things that animals have learned for themselves and things that they have learned by copying other individuals. The use of medicinal plants, for example, may be the result of each individual tasting one of the plants out of curiosity and finding that it sometimes felt a great deal better for having done so. There need be no knowledge community involved in all this – merely the accidental discovery by a set of individuals, each acting on its own. The widespread occurrence of parasitical conditions and the availability of many suitable plants with anti-parasite properties means that it is difficult to distinguish between individual learning and cultural transmission.

Of one thing, we can be sure, however, and this is that chimpanzee and human herbal medicine of this kind differ in one crucial respect: there is no magical element to chimpanzee folk medicine. Human folk medicine is almost always attended by associations of the supernatural: they are often prescribed by shamans or witchdoctors, who invoke magical powers from a world we cannot see, practise special rituals and make invocations to invisible spirits in order to ensure the efficacy of the herbs they give us. Of this, there is no sign at all in chimpanzee folk medicine. What they do seems to be an entirely private business.

But, bearing in mind the concerns we had earlier about placing too much emphasis on the role of language, should we insist that the real marker for culture ought to be the fact that one individual acquires a habit from another by some form of social learning, perhaps without realising at first that there are benefits to be had by doing so?

The uncertainty on this point has led the psychologists, in particular, to place a much stronger emphasis on the psychological mechanisms by which individuals acquire behaviour patterns. Their argument is that we can only really be sure that we are dealing with culturally inherited behaviour when we discover actions or beliefs (a) that bear no relationship to any ecological or other biological problems of daily life (and so cannot have been learned by trial and error in response to a particular environmental problem), and (b) which are copied mindlessly from another individual *just because* the source individual does it that way. Something as arbitrary as this simply cannot be anything except a bona fide case of culture. This does not exclude the possibility that many other kinds of behaviour with more practical benefits might not also be culturally transmitted. But the problem with all these examples, they would argue, is that we just cannot be sure whether the behaviour was acquired by individual trial-and-error learning or by blind social transmission (imitation). It is the latter kind of behaviour that we need to concentrate on in our search for evidence of culture in other non-human species.

In Search of a Holy Grail

In the past, much has been made of cultural behaviour in animals. The opening of the old-fashioned cardboard milk bottle tops by English tits during the 1950s and the habit of washing sweet potatoes by Japanese macaques have been famously touted as examples of culture in animals because they seem to involve the social transmission of behaviour. The important issue here has been the claim that these behaviours have been passed on from one individual to another by observational learning: an individual sees another behave in a particular way and copies it. Aren't these examples of just what we mean by cultural behaviour in humans?

In recent years, however, considerable doubt has been cast on these claims. Not that Japanese macaques and blue tits did *not* perform these actions – nor that their acquisition by subsequent generations was not a social process. These claims are beyond doubt. What has been brought into doubt is whether these are cases of genuine *copying* – of social *imitation*. After carefully analysing what happens during social learning, comparative psychologists have come to the conclusion that social learning is not a simple phenomenon underpinned by a single learning mechanism. Rather, they distinguish at least three quite separate mechanisms that can give rise to social learning. These are stimulus enhancement, emulation and imitation.

The differences between these three mechanisms hinge on just what is transmitted or passed on between model and imitator. The process of stimulus enhancement focuses on the problem that the animal needs to solve. In other words, the imitator's attention is drawn to a problem in the physical world by another's behaviour, but it works out the solution for itself by trial-and-error learning. One particularly inventive tit used conventional trial and error to discover that the cardboard tops which fitted inside the top of 1950s milk bottles could be levered out, so giving it access to the cream that floated to the top of the bottle. Its companions subsequently noticed it perched on a milk bottle top, and joined it. They learned that there was a nutritious source of food in these bottles. Later, these birds tried alighting on other milk bottles, but found the caps in place and so were unable to get at the cream. In due course, they too learned by a process of trial and error how to get the caps off the bottles for themselves. And so the habit spread slowly through the population. However, and here is the real rub, the rate with which the habit spread through the population really was very slow, much slower than we would expect if the birds were copying (i.e. imitating) the behaviour of their companions rather than having to work it all out for themselves. And it was really the slowness with which the habit spread that gave the game away. A process of imitation should have resulted in a much faster rate of spread.

The second mechanism, emulation, is rather like the converse of stimulus enhancement. In this case, the imitator observes the behaviour of the model and uses that as a guide to identify a problem that is worth solving. For example, the Japanese macaque companions of the young female Imo, the first member of the group to engage in potato washing, saw her doing something with a sweet potato in the sea and were led to investigate what she was up to and to try putting potatoes in the sea for themselves. Having discovered that the water washed the gritty sand off the potato and made it more palatable (perhaps because the sea salt enhanced its flavour), they then proceeded to repeat the behaviour. They did not copy Imo's behaviour so much as use her as a guide to discover an interesting feature of the world. Once again, the slowness with which the behaviour spread points to a trial-and-error learning process rather than straightforward copying: on average, it took each animal in

Imo's troop of Japanese macaques two whole years to learn to potato-wash. This slowness makes it seem unlikely that they were simply imitating what she and others did. But it makes more sense if what they were actually doing was figuring out the solution to the problem for themselves.

In true imitation, by contrast, the imitator simply copies what it sees without necessarily thinking about the point of the exercise. It is the equivalent in humans of adopting a fashion simply because others have. We do not ask why or how our model is doing what he or she does, we simply do it. So while wearing hats is done because it is useful, wearing baseball caps backwards is done simply because we've seen someone else wearing them that way.

And it seems that humans – and particularly young children – are especially good at imitation. Several studies have attempted to teach young apes and children to do a particular task, such as how to open a box to get a food reward. However, some of the apes and children are taught to open the box in an obvious and easy way, and the rest in another more difficult way. The children happily copied whichever way they were taught. The apes, in contrast, more often than not tended to open the box in the easiest way they could rather than just copying what their human model had demonstrated. It is the speed and facility with which human children learn to copy what they are shown that is so impressive. In contrast, the whole process is more laborious for apes. Yes, they may eventually learn to copy what they have been shown, but it can take a long time for them to reach perfection. In contrast, children often need only one or two exposures, and they have it perfectly.

Still, we need to be careful not to over-interpret these differences. Chimpanzees tested on the copying routine do occasionally seem to have copied what they've been shown. And there is circumstantial evidence from the wild to suggest that imitation may occur. One example of this is the difference in techniques used to dip for ants between Gombe in East Africa and the Taï forest in West Africa. In the former, the chimpanzees use a long stick (on average, about two foot long) and allow the ants to climb half way up the stick before scooping them off between finger and thumb, repeating this routine at about 2.6 times a minute. In the Taï, the chimpanzees use a shorter stick (on average, a foot long), allow the ants to climb only a third of the way up the stick, and then draw the stick through their lips to remove the ants, repeating this manoeuvre as often as twelve times a minute. However, because the Taï technique gives many fewer ants per manoeuvre, the end result is that the Gombe chimps do much better – they can average around 760 ants per minute when ant-dipping, as against a mere 180 ants per minute for the Taï chimpanzees. If trial-and-error learning was part of the process, we could reasonably expect the Taï chimps to have figured out a more efficient way. The fact that they have not suggests that they may have learned the technique by copying rather than by emulation.

Psychologists are, of course, rightly nervous of relying unquestioningly on casual observations from the wild. We know nothing of the history of these two behaviour patterns. In such uncontrolled situations, we are especially prone to being unwittingly taken in by the Clever Hans effects. We often have little idea of the background to the behaviours in question, and can say nothing about the possible learning history involved. It is just possible that the behaviours of West African safari ants are sufficiently different from those of the East African ants that the Gombe technique might prove to be an unmitigated disaster if it was applied in Taï. We simply do not know. There are just too many unknowns and imponderables. Hence, psychologists prefer carefully controlled experiments in the laboratory, contrived though they may sometimes seem to be.

There is one feature of social learning that we have not yet considered, and this is teaching. We might see teaching (as opposed to imitation by copying) as guided learning. In humans, the teacher monitors the actions of the pupil and corrects or guides the pupil when a mistake is made. This is a very special process that is clearly characteristic of much of what we as humans do. Teaching allows us to learn rapidly – and with the fewest possible mistakes – many of those features of human culture that define us and our societies. However, we should be cautious about placing too much emphasis on teaching as a learning process when trying to understand the differences between us and our nearest relatives. After all, teaching is a process for speeding up the learning process; it is not the learning process itself. The pupil essentially still learns by imitation, copying or trial and error. Nonetheless, we can legitimately ask whether teaching is common among our primate relatives or unique to us.

The short answer to this question seems to be that, by comparison with what we see in humans, teaching is extremely rare in the animal kingdom. To be sure there are documented examples of what looks very much like teaching. Cat mothers often bring half-alive mice or birds for their kittens to practise killing. Chimpanzee mothers leave intact nuts for their offspring to crack. However, we perhaps need to recognise here that there is a crucial distinction between *facilitating* (providing an opportunity for trial-and-error learning to occur) and *teaching* (deliberately showing how something should be done). Christophe Boesch, who has studied the Taï chimpanzees for two decades or more, was able to document many cases of the former, but only two apparent instances of the latter (and many would regard even these as somewhat charitable interpretations). In one case, a mother slowed down and modified her nut cracking so her offspring could follow it more easily; in the other, a mother altered the position of her son's nut when he was having trouble hitting it. But, it takes on average about ten years for a chimpanzee to learn how to crack nuts efficiently. In contrast, it takes only a few weeks of intensive training for a child to learn how to tie its shoe laces, a behaviour that is considerably more complex. As with the blue tits and the Japanese macaques, this rather suggests that learning how to crack palm nuts is really based on simple stimulus enhancement or emulation, combined with trial-and-error learning, not on imitation or teaching.

In humans, the key to teaching, as Mike Tomasello has pointed out, is intention: the teacher deliberately intends to modify the actions of the pupil. Does either of the Taï chimpanzee examples meet this criterion? The answer is: (probably) yes. But even so, the fact that we can find only two clear examples of teaching in literally thousands of chimpanzee-years of observation by scientists in different parts of Africa speaks volumes. Clearly, the capacity exists at least in minimal form, but it is not used as often as we might expect were we studying humans. In humans, teaching goes on endlessly, day in day out.

And here, perhaps, is the unexpected lesson of the comparative work on cognitive development in humans and apes. Human babies are imitation machines who seem to suck up anything and everything they come across that involves imitation of another individual's behaviour. Teaching helps to guide that up-take, but without the human child's seemingly infinite capacity for imitation, it is doubtful whether any amount of teaching by the parent would help in the absorption of so much behaviour in so short a space of time. In contrast, young chimpanzees seem more proactive and clued in to finding things out for themselves. This contrast is a bit of a puzzle really, because, when you think about it, blind copying would seem to be a remarkably dull-witted intellectual capacity. It is hardly rocket science. Why should the young of the most cognitively advanced species on earth seem to be less clued into the world in which it lives than the young of other intellectually less well endowed apes?

A Very Cultured Ape

Impressive as these examples of chimpanzee culture are – and I really have no problem about using the term 'culture' when talking about them – they remain, in the final analysis, unsatisfying. For one thing, it is the relative scarcity of genuinely cultural behaviour that is troubling. A grand total of 39 elements of cultural behaviour from the hundreds of thousands of hours of chimpanzee-observation in the wild and in captivity is not an impressive tally. Were we to conduct the same experiment on humans, we would surely find so many examples of differences that the chimpanzee catalogue would fade into insignificance, even if it was ten times as large as it currently is. But, there is another and more troubling absence in the great ape story: we still see nothing that smacks of those activities that form so fundamental a part of human culture – story-telling and music, and beyond them that whole panoply of religion and ritual, and the significance of a spirit world that sits apart from the real world in which we all live.

The more I think about it, the more obvious it becomes that there is one fundamental reason why chimpanzees will never write the tales of Shakespeare or compose the poems of Baudelaire or T. S. Eliot. The simple fact is that they do not have the levels of intentionality to be able to do it. Even if great apes *could* aspire to theory of mind (second-order intentionality) that would not grant them the capacity to produce these most human of all cultural phenomena. When Shakespeare wrote *Twelfth Night*, he *intended* [1] that his audience should *realise*

[2] that the much derided Malvolio *believed* [3] that his mistress Olivia *wanted* [4] to marry him instead of his being her servant (with the levels of intentionality once again marked in numerical order). And in writing *Othello*, he *intended* [1] that his audience *realise* [2] that the eponymous moor *believed* [3] that his servant Iago was being honest when he claimed to *know* [4] that his beloved Desdemona *loved* [5] Cassio. Shakespeare's literary efforts were a fourth- or even fifth-order task – and fifth-order tasks, as we saw in Chapter 3, are exacting and challenging even for humans of above average intelligence. Even if a chimpanzee could speak, it would not be able to follow the convolutions of the plot – despite the fact that, to do so, it would need one level of intentionality less than the great Bard had required to write it. There is no evidence to suggest that the more advanced levels of intentionality above second-order are anything other than a purely human preserve.

In short, while language itself is not essential to literature, advanced theory of mind is. Language is, of course, essential for *transmitting* the story from one individual (the composer) to another (the listener), otherwise we would never get to hear about it. But language is not essential for *constructing* the story in our minds. We can each compose a grand Shakespearian tragedy in our heads, and go on to enjoy the artistic devices and cleverly constructed plots without ever having to speak a word. We do not even have to do it in words. I could compose all the stories in the universe in what is sometimes referred to as the 'language of thought' (the silent thoughts that go on inside our heads in a kind of wordless, perhaps even visual, stream). It would be tough on the rest of you to have missed out on all my works of undoubted genius . . . but that, as they say, is not my problem. It would be much like Samuel Pepys writing his diaries in his own secret code for his own enjoyment when he read them again later.

Nonetheless, there is something unsatisfactory about suggesting that my silent literary endeavours constitute high culture. Without a community of story-tellers and their audience, it is questionable as to whether we would really have Culture with a capital 'C'. The cultural community that make sense of the stories I tell interpret them in terms of their own individual experiences of life, applaud the good ones and scoff at the bad, adding their own nuances of interpretation as they do so. This, surely, is what human culture is all about. Story-telling becomes culture because the stories we tell come to influence the minds of others. Strictly speaking, we do not need language for that, but we do need some form of communication. Mime might suffice, Egyptian hieroglyphs would do admirably. Language does happen to work particularly well, however.

If literature remains a purely human domain because of our advanced mind-reading abilities, then does it provide us with any purchase on the question of when this key defining feature of humanity evolved? The short answer is no. Because stories do not fossilise. But there is one form of story-telling that does leave its imprint in the archaeological record – religion. Religion requires us to be able to conceive of imaginary worlds, worlds that we do not directly experience. We have to be able to step back from the immediacy of our everyday experiences and ask: 'Could the world be otherwise than as I experience it? Could there be a parallel world inhabited by beings that I cannot see and touch directly in the way I see and touch objects and individuals in the world we all inhabit?' We have to be able to imagine that the world is other than it seems to be from our everyday physical experiences of it, and to be able to suppose that this parallel universe out there somewhere, peopled as we imagine with other beings, can influence our world – and perhaps, in their turn, be influenced by us.

*

Herein, then, lies the great divide between ourselves and our ape cousins: the world of the imagination. We can imagine that something can be other than it is. We can pretend that there are fairies at the bottom of the garden. We can construct rituals and beliefs that have no intrinsic reality other than in our heads. Other animals cannot do that because they cannot step back from the world and wonder how it might be if it was different from the way they perceive it to be. And so we are brought full force to the one thing we have skirted carefully around, the issue of religious belief.

7 Thus Spake Zarathustra

The dancing had continued late into the night.

When he awoke the next morning, he felt invigorated by the prospect of what he knew lay ahead. The men would gather early, down by the riverbank. Then, silently and without any discussion, they would set off along the trail that led through the woods up the valley. An hour later, and they would be at the cave. They would take the carved stone lamps from a niche on one wall, and light them carefully with a wick of glowing tinder brought from the campfire. The lights would flicker up uncertainly, casting their dull glow into the darkness of the cave ahead of them, testing apprehensively for the presence of bears or wolves.

It was some hours later that reality caught up with his thoughts. But now they were there in the cave mouth, preparing for the slow, measured journey down the dark passages. As they went deeper, the air became damp and he could taste the mustiness; here and there, water oozed from the rock and ran down the wall to form puddles and rivulets on the uneven floor. The men stepped carefully across the wet surfaces, and edged their way on into the dark ahead.

Eventually, the passage opened out into a large cavern. The light of their lamps did not reach the roof, but the echoes of their whispers told them that the roof was high and the space it formed *large. Those carrying the lamps took them to the centre of the cavern and laid them in a tight circle on the rough rock floor. Silently, they rejoined the others waiting at the cavern's edge.*

They were waiting for the oldest among them to take the lead, as he always must. But the old man had sat down on the rock floor and seemed deep in thought. The artist shivered in the coldness. He knew that finding the right moment was important, so like the others he waited patiently. And so they continued to wait, the stillness broken only by the occasional scuff of movement when one of the men changed position awkwardly.

Eventually, after what seemed an eternity, the old man rose to his feet and walked out into the centre of the cavern's space. The others silently followed and gathered in a circle round him. The old man began to hum. The notes of the cave song floated falteringly into the emptiness in his rich bass voice, as yet unbroken by age. He began to move in a wide circle round the lamps, his limbs moving jerkily with a stamp on each footfall in the way they had danced around the campfire with the women the night before. The others began to join in the song of the cave, one by one. As the volume grew, the notes reverberated from the cavern walls, tumbling and cascading back and forth, the lighter voices mixing with the heavier bass in a counterpoint so ethereal it set the senses on edge. One by one, the men tucked into line behind the old man, their bodies flowing into the rhythm of his dance as he led them round the lamps.

After a while, the dance began to develop a new momentum, steadily building a faster beat on each successive round as the men dipped and swayed into the rhythm of the song. Sweat began to pour, and the cavern suddenly seemed to grow warm. The effort of the dancing made it difficult to keep singing, but the artist knew he must. And gradually, with an inevitability that spoke of deep mysteries, the dance and the song began to work their magic. He knew the moment would come soon.

And then it did. The first sign was the heaviness of the liquid dripping onto his chest, too heavy and sticky to be sweat. Blood was dripping from his nose. And then his head seemed to burst inside him, the intensity and whiteness of the light catching him by surprise as it always did, speeding him into a rush of intense joy. And then he was away. He could feel himself floating gracefully out from his body as it slumped on the cavern floor, the lights of the other world flashing in starbursts around him.

He saw, looming out of the dark beyond, a tall-necked creature, half-man, half-deer. It cantered over to him and gently began to nuzzle him inquisitively. He was always there at the start, his spirit guide. Grasping the beast's mane, the artist turned and pointed the way into the spirit world. And away they sped, effortlessly gliding across the landscape, searching for the secrets of the world beyond.

Religion would seem to be a truly universal trait among humans. Every human tribe that has ever been encountered has some form of belief in a spirit world and most (but maybe not all) have some sense of an afterlife. All engage in rituals and prayers intended to placate, cajole or entice the denizens of that unseen world to look favourably on the poor long-suffering members of the human race. At the same time, we have no evidence of any kind that would seriously suggest that any other species aside from ourselves has anything remotely resembling religion. This is not simply because other species lack language. Language is important in formalising religion within a community, in allowing us to agree on the nature of the gods in whom we believe and the afterlife that we hanker for. But this is not what makes religion or religious belief in the individual possible.

There are three quite distinct questions we can ask about human religious experience: (1) Why are we the only species to have religion and believe in a parallel world? (2) What function did religion serve for our ancestors, and to what extent does it still serve that function for us today? (3) When did religion first appear in human history?

An Edifice for My Father's House

Even a cursory glance around the world's myriad religions should convince us of one thing, and this is that religion serves several different, but often equally important, purposes in the lives of recent and modern humans. These functions would seem to be: (1) providing coherence for the world in which we live (a metaphysical scheme that explains why the world is as it is, and thus makes sense of it for us); (2) allowing us to feel we have greater control (through prayer and other rituals) over the vagaries of life than we would otherwise do; (3) enforcing rules about how we should behave in society (ethics and moral systems); and (4) allowing a minority to exert political control over the community.

I detect in these two quite separate agendas. One seems to be associated with trying to allow us to cope with a world that is not always as benevolent towards us as we might wish. The other seems to have much more to do with social control in a very broad sense.

The idea that religion provides a unifying framework for the world in which we live is hardly new. Sigmund Freud, among others, supposed that religion fulfils the role of science in primitive societies. Although contemporary anthropologists have often vigorously resisted such an interpretation (mainly because they have a deep antipathy towards anything that might be taken to suggest that tribal societies are primitive, second-class or intellectually inferior to western 'scientific' ones), there is a great deal to be said for this view. One thing that most religions often explicitly do is to provide accounts of how the world came to be and how it works. These accounts (think of the creation stories in the Bible or the Rainbow Snake myths of the Australian Aborigines) often provide us with an explanation as to why the world works the way it does and, in doing so, they invariably tell us what our role in the world is. This is not to suggest that quasi-religious explanations of this kind are wrong-headed, or should be seen as somehow second-class because they are at odds with what we now know from science. That is to miss the whole point of the exercise. The purpose of any such system (and, at this particular level, this applies to science as much as to religion) is to provide coherence to an otherwise rather confusing world.

Humans are extremely good at recognising correlations in the world. That is what allows them to function so effectively at the ecological level, whether as hunter-gatherers, pastoralists or horticulturalists. As I have detailed in my book *The Trouble with Science*, traditional societies are often able to produce phenomenological accounts of the world that are at least as good as those of modern science, and they are able to use that information very effectively in managing their economic activities. This should not surprise us, given that we are all observing the same world. However, there is a fundamental difference, as we noted in Chapter 3, between knowing *that* something is the case in the world and knowing just *why* it has to be that way. The latter often has to do with the hidden structures of the way the world is constructed (the laws of physics, chemistry and biology) and is often quite opaque to us unless we have the sophisticated tools of modern science that allow us to probe beneath the surface appearance. Probing beneath the surface – which is what science busies itself with – is all very interesting, but it does not necessarily help us survive any better. Indeed, it can have the opposite effect. As the San peoples of southern Africa observe, there is no point is entering the lion's den just to see whether lion cubs' eyes are open at birth: useless knowledge costs lives. Given the rather more pressing concerns they have over day-to-day survival, the San hunter-gatherers would no doubt regard science in much the same way they would knowing whether the lion cub's eyes are open – an interesting, but rather trivial and pointless activity.

However, the trouble with knowledge in this form is that the human mind would soon be overwhelmed by the mass of apparently random correlations it is capable of divining in the natural world. Having a schema that makes sense of at least some of these greatly reduces the cognitive load: less has to be remembered because many of the bits and pieces can be inferred from a few basic principles that provide explanatory coherence. The point here is that it does not really matter what that schema is or how well it reflects the true underlying reality of the phenomenal world. To do the job, it simply has to make sense of the world we experience, by linking the apparently unrelated bits and pieces in a logical and internally (even if not externally) consistent way. And in some respects, the simpler and more easily understood the grand scheme, the better it is. There is little to be gained by having an explanation that is so complex or difficult to confirm that we waste valuable time on it when we could be out foraging or finding mates.

If our view of the world is a reflection of how our societies happen to be organised, as some anthropologists have argued, then that is fine so long as it provides us with a useful basis for organising our knowledge. Unlike science, religion is not necessarily intended to give us the exact answer, merely one that works for most everyday purposes. Of course, it remains true that the better our theories reflect the true underlying reality of the world, the better they will work (in terms of allowing us to predict or control the future) and the more successful we will be in consequence in our day-to-day survival. But, as is so often the case in real life, the law of diminishing returns means that there will always be a point after which it is just not worth investing more time and effort into figuring out the underlying reality. In traditional societies, anything that does the trick will do.

Theory of mind, and the more advanced forms of cognition this underpins, allows us to step back from the brink of the world and ask why it has to be thus. Without that ability, we would not be able to do science as successfully as we do, for science requires us to be able to ask whether it could have been otherwise, to imagine something else behind or within it. Only by taking that step can we then ask why it has to be the way it is – or, indeed, whether there is anything we can do to change it. Animals, even those as sophisticated as chimpanzees, cannot do that. Without the higher orders of intentionality, their noses are, as it were, thrust firmly up against the grindstone of experience: they cannot step back far enough to see it other than as their senses tell them it is.

As beneficial as this faculty of ours is, it does not come cost-free. The cost is that we are very soon forced to confront the uncomfortable fact that the world is not the easiest place in which to survive. It constantly throws at us events and circumstances that are beyond our control. We are overwhelmed by flash floods or rampaging elephants, our villages are sacked and our storehouses ransacked by human marauders from over the hill, diseases strike down our children without warning. With minds and souls as sensitive as ours, these are not events that we can easily take in our stride: the pain of losing loved ones is always intolerable. We need something to bolster us against the onslaught, to hold our spirits up just long enough to tide us over the disaster period and into the better times that must surely lie ahead. Were this not so, we would all rapidly succumb to the inevitable despondency and despair, and end by giving up on life (as some who get trapped in the depths of despair occasionally do).

A metaphysical system and the power of prayer provided by religion combine to give us just enough strength to see us through these depths and out the other side. It is a commonplace that religious people seem, in general at least, to be happier than those who lack religious belief. Indeed, there is good empirical evidence to support this. Religious people in general do suffer less frequently than non-religious folk from both physical and mental disease; moreover, when they do go down with something, religious people recover more rapidly from both the disease and any invasive treatment (such as a surgical operation or chemotherapy). One reason must be that they feel they have greater control over the circumstances that beset them – God will look after them, whatever happens. The Islamic expression *Inshallah* ('God willing') is less fatalistic than Christians sometimes assume it to be: its sense is more that God knows what he is doing, and he watches over his own. It is surely no accident that almost every religion promises its adherents that they – and they alone – are the 'chosen of god', guaranteed salvation no matter what, assured that the almighty (or whatever form the gods take) will assist them through their current difficulties if the right rituals and prayers are performed. This undoubtedly introduces a profound sense of comfort in the face of adversity.

An Opium for the People

There is, in addition, a rather more prosaically biological side to religious experience that has emerged only during the last decade. Many of the practices that religions enjoin on their followers are just the kinds of activities that are likely to be good at stimulating the production of endorphins in the brain. Different religious traditions have, of course, placed different emphases on the kinds of activities that are deemed appropriate for religious observance, but it is striking that so many have placed so much emphasis on the infliction of physical pain and/or stress. These have included fasting, dancing or other rhythmic movements (think of the rhythmic bobbing of orthodox Jews praying at the Temple wall in Jerusalem, the repetitive counting of rosary beads and similar prayer devices), flagellation and the painful tasks imposed on pilgrims (such as walking the Stations of the Cross on one's knees or, in the Buddhist and yogic traditions, long periods sitting motionless in positions that are difficult to adopt), painful or stressful initiation rites in many tribal societies, communal singing (especially the tonally deep sustained forms that are typical of chanting, but also the lusty singing of hymns in the more evangelical traditions of Christianity), the intense rhythmically repetitive singing of the *qawwali* tradition in Sufi Islam, the long hours spent locked in services, the emotional rollercoaster induced by all the best charismatic preachers . . . The list could go on and on.

All these practices impose low but persistent levels of stress on the body, and it is precisely this kind of persistent low-level stress that is particularly effective at stimulating the production of endorphins. Unlike the neural pain-control systems (which are designed to cope with the intense kinds of sharp pain induced by actual injury), the endorphin system is specifically designed to allow us to cope with those kinds of discomfort that come from long-running stresses on the body. Marathon runners, for example, have a great deal to be thankful to the endorphin system for, since it is this that keeps them going through the pain and grind. Indeed, it is probably the endorphin system finally kicking in that is responsible for the 'second wind' phenomenon so familiar to long-distance runners. Regular joggers, whose insistence on a daily dose of body-bashing puts their muscles under light but repeated stress, will be familiar with their personal endorphin systems too. The frequency with which they stimulate it is such that they become addicted to jogging; when they do not get their daily jog, they experience all the usual kinds of cold turkey (albeit in mild form) that drug addicts go through when deprived of opiates – the edginess and irritability, the inability to settle until they have had their daily 'shot'.

Religious practices seem as though they are purpose-designed to give us that opioid kick that makes us feel so much better able to cope with the vagaries of the world and, perhaps just as important, so much more at peace with our neighbours. But it may also be the case that the endorphin production associated with these practices stimulates the immune system into greater activity, thus directly protecting the body against disease and injury. In fact, there's an interesting parallel here with the pacing often found in caged animals in zoos. Once thought to be a sign of boredom, it has now been shown to stimulate the production of endorphins. For better or for worse, it probably helps the animals cope better with the stress of confinement.

So pervasive are practices of this kind that they have sometimes come to provide the central plank on which particular sects have established their identities. The most famous of these is undoubtedly the Flagellants, a movement that was born in 1260 in the region of Perugia in Italy. Bands of 50–500 penitents marched from one village or town to another, pausing at each church to whip themselves with scourges in a carefully orchestrated and highly charged ceremony that attracted enormous interest as a public spectacle, and which often resulted in rich and poor alike streaming to join. The infliction of what was often severe pain and even injury inevitably meant that the movement had a relatively short lifespan. Nonetheless, it experienced a major revival a century later as the Black Death swept through Europe in 1347–8 in what became a desperate but misguided attempt to banish the sickness by mass atonement for the sins that were assumed to have brought it – at least until it became apparent that the itinerant bands of penitents were actually helping to spread the plague from one town or village to another, at which point towns started to bar their gates against them.

In Orthodox Russia, the Khlysty ('flagellators') and the Skoptzy ('mutilators') sects aimed to achieve a state of religious ecstasy through self-imposed physical pain. Perhaps because the Skoptzy advocated self-immolation (or, in the case of women members, removal of the breasts), this particular offshoot was short-lived. But the Khlysty movement had a long history: having first emerged perhaps as early as the 1360s, they were still in existence as a semi-heretical sect within the Orthodox Church late enough for the 'mad monk' Rasputin (he of the downfall of the Romanovs fame) to come across them when he visited the monastery of Verkhoture in the 1890s. Nor are these practices confined to Christendom. Islam has its own versions in the dervish sects of the Middle East. The annual Shia rituals in honour of the martyrdom of the Imam Husain and his family at Karbala are a potent example of this tradition.* During the celebrations, lines of men rhythmically slash at their chests with knives or flagellate their bare backs with heavy whips until the blood runs, while the accompanying women weep and howl in memory of the terrible fate that befell Husain in 680 AD.

In many respects, however, these opioid effects are but the consolation prize for the masses. The real adepts have even more profound benefits to gain. Barely a decade ago, Andrew

* Husain was the second son of the caliph Ali (son of the Prophet Mohammed), whom the Shia regard as the rightful heir to the Prophet. Husain inherited the caliphate on the death of his older brother, Hasan, but his claim was disputed and he and a party of companions were later massacred.

Newberg (a neuroscientist) and Eugene d'Aquili (an anthropologist) discovered that individuals who can achieve a heightened state of religious ecstasy (such as that achieved at the endpoint of meditation) exhibit very specific patterns of brain activation. Brain scans of individuals in this state have a greatly reduced level of activity in an area in the posterior parietal lobe of the left hemisphere (the area mainly responsible for our sense of spatial self) – and, incidentally, a great deal of generalised activity in the right hemisphere, though they made rather less of this. On the basis of this evidence, they have argued that carefully orchestrated mental practices (techniques developed by mystics in all religions) allow adepts to disengage a bundle of neurones in the posterior part of the left parietal lobe of the brain (roughly above and behind the left ear). Once these neurones are disengaged from the control of the rest of the brain, they release a series of impulses down through the limbic system to the hypothalamus, which then sets up a feedback loop between itself, the attention areas in the frontal cortex (which have been responsible for blocking off the parietal lobe neurone bundles) and the parietal lobe itself. As this cycle builds, it leads to the complete shut-down of the spatial awareness bundles, generating as it does so a burst of ecstatic liberation in which we seem to be united with the Infinity of Being, often in a flash of blinding light. For obvious reasons, this bundle of neurones in the parietal lobe has been termed the 'god spot'.

However, the explanation for this effect may not be quite what Newberg and d'Aquili suppose. The clue lies in the fact that the hypothalamus is involved in the circuit. The hypothalamus just happens to be an area of the brain that is particularly prominent in the opioid story: it is a major site from which endorphins are released into the brain. That burst of peaceful nothingness that comes at the point of meditation may be nothing more than a familiar opioid surge. The important point about their discoveries, however, is that these effects can be generated by mental self-stimulation by adepts. Interestingly, the experience of the mystic at this point (the suffusion of the mind with a blinding light, the sense of being at peace and at one with God, the semblance of the mind or soul leaving the body and hovering above it) are identical to those that occur in near-death experiences. The latter are thought to be the result of oxygen starvation to the brain. One likely explanation for the mystical version of these experiences is that adepts have discovered ways to induce oxygen starvation to the brain, perhaps even selectively to particular key areas within the brain.

In short, mystics have found the secret of the universe. Contrary to Douglas Adams in *The Hitchhiker's Guide to the Galaxy*, it is not the number 42 but rather the ability to self-induce an endorphin surge. The rest of us mere mortals have to make do with more prosaic forms of physical stimulation to elicit the same kinds of effect at much lower levels of intensity.

Communality and the Sense of Belonging

Important as these psycho-pharmacological factors are, there are other important benefits to be derived from being involved in religious movements. People who belong to organised religious groups are also members of a community and that community may be mutually supporting in a particularly intense way. They feel that they *belong*. There is a considerable volume of evidence to suggest that people's ability to resist disease and to cope with life's many traumas is directly affected by the size of their social network. A large study carried out during the 1950s in Newcastle, England, showed that, even in modern industrial societies like ours in Britain, children from larger extended families suffer fewer ailments and die less often than those from smaller ones. Similar results have been reported from a study of a rural agricultural community in Dominica in the Caribbean.

These, of course, are statistical effects: it is not that everyone in a large family has an easier time of it, merely that on average they do better. But the effects are there and quite robustly so. Mortality rates among the early European settlers in North America were also related to the size of the kinship group. In the famous settlement that was established at Plymouth, Virginia, by the *Mayflower* colonists in 1620, mortality rates over the first winter were significantly higher among those who came alone than those who arrived in family groups. The mean relatedness* to all other members of the colony was about 0.8 for those who survived, but only about 0.2 for those who died during that first terrible winter.

A similar story emerges from the famous Donner Party incident, one of the legendary events in American folk history. The Donner party consisted of eighty-seven men, women and children who set off in April 1846 in twenty covered wagons from Springfield, Illinois, to start a new life in California. Thanks to a series of unfortunate delays along the way, they reached the passes of the high Sierra Nevada mountains much later than they had intended, and October found them snowed in at high altitude. Unable to proceed or retreat, they battened down as best they could to face the winter. By the time the thaw set in the

---

* In this study, relatedness was measured as the sum of all the degrees of relationship between the individual in question and everyone else in the colony: we are related to our parents, children and full siblings by 0.5, to our grandchildren, grandparents, aunts/uncles, nieces/nephews and half-sibs by 0.25 and to our cousins by 0.125, and so on. These values (known as the *coefficient of relationship*) reflect the probability of sharing a particular gene by descent from a common ancestor, given that we inherit half our genes from each of our parents (or, alternatively, each parent passes only half of its genes on to each of its offspring). The coefficient of relationship forms the cornerstone for one of the central pillars of modern evolutionary biology, the theory of kin selection – the mechanism that predisposes us to be especially generous towards relatives. For more explanation on this, consult any textbook on behavioural ecology or evolutionary psychology.

following April, forty (nearly half) of the original party had died as a result of the appalling conditions they had been forced to endure. But a disproportionate number of those who died were strapping young men travelling on their own and a disproportionate number of those who survived were members of family groups. Adult males who survived were accompanied by an average of 8.4 other family members, whereas the males who succumbed were travelling with an average of 5.4 other individuals. Only three of the fifteen single men that had begun the trek in Springfield made it all the way to California; the only woman who died was travelling in a group of four, compared to an average of ten individuals in the families of the women who survived.

There is something about families and the sense of belongingness that they engender that bucks our spirits, makes us generally feel better able to take on the world and genuinely do better at it. In fact, there is some evidence that a strong support network has a positive effect on our immune systems – the very thing that allows us to resist disease and cope better with life's vagaries. The UK national health statistics, for example, suggest that the frequencies of psychological depression are directly related to the breakdown of kin support networks. The role that religion may play in providing that same sense of community and belongingness is surely obvious. It cannot be an accident that, at least in the Christian tradition, the church is referred to as a family; the whole iconography of God the Father and Mary the Mother (of God), the use of 'father' as an honorific title for priests, not to mention the liberal use of the terms 'brother' and 'sister' to refer to members of the congregation, resonate with this same sense of family.

What all this suggests is that a religious sense may have evolved to facilitate the psychological bonding of the extended family groupings in which we have spent most of our evolutionary history. It was there to provide a sense of community and belonging. Sharing a common viewpoint and – however arbitrary they may be – a common set of dietary laws, rituals and behavioural prohibitions may be as important a badge of group membership as sharing the same dialect. Indeed, the more demanding the practices the better they are as evidence of commitment to the common ideal. That sense of community acquires a particularly heady intensity when it is focused on some particularly charismatic figure. Under these circumstances, our willingness to sublimate our personal desires to the common will makes for an explosive mix.

The Communal Bond

On 18 November 1978, at the behest of the Reverend Jim Jones, 923 men, women and children committed mass suicide (some of them, albeit, less than voluntarily) at the People's Temple in Jonestown, Guyana, on the northern shore of South America. Jones – who at one point claimed to *be* God – had taken his followers to Guyana from the USA because he believed that the USA was about to be engulfed by Armageddon, although a more likely reason was that he was beginning to come increasingly to the notice of the FBI. It was a fact-finding visit to Jonestown by a US congressman (and his subsequent murder, apparently by Jones's henchmen) that sparked the final denouement for the People's Temple movement. The pressure was on and it may be that Jones could see no other way out of the closing net.

Such events are by no means uncommon. In late February 1993, David Koresh and at least seventy-three members of his church, the Branch Davidian, effectively committed what amounts to mass suicide during the course of an attack by US security forces on his fortified base, Mount Carmel, at Waco,

Texas. As the decade wore on, at least three other groups of individuals belonging to various obscure latter-day religious or New Age cults committed voluntary suicide in California, Switzerland and Canada.

What is it that allows one man (and in most, though not all, cases it is a man) to persuade others to follow him to the brink and beyond? Were these isolated cases, we might perhaps satisfy ourselves with explanations in terms of the madness of the few, or the self-destructiveness of a Joan of Arc. But the momentary dramas of such events are merely the tip of a much larger iceberg. Men and women through the ages have been willing to follow almost any Pied Piper that has chanced to come their way. The histories of the three great religions of the western world – Judaism, Christianity and Islam – are littered with such cases.

There had been a long string of Jewish prophets and messiahs in the century or so prior to the time of Jesus of Nazareth and an equally long list of now largely forgotten names after his time. Among the latter were Simon Bar Kokhba (although some argue that this early-second-century AD guerrilla leader never made any messianic claims), Moses of Crete (fifth century AD) and Sabbatai Zevi (1626–76), each of whom attracted a significant band of followers and held influence over a wide area in the eastern Mediterranean and beyond at the height of his fame. Sabbatai Zevi is widely regarded as the last of the great Jewish mystics and prophets. Encouraged to declare himself the messiah, he attracted widespread fame among the Jewish communities throughout Europe and Asia Minor during the middle decades of the seventeenth century. Although he converted to Islam to avoid execution at the hands of the Turkish sultan after being captured in Constantinople in 1666, his fame long outlived his death in exile. There was at least one attempt to revive the sect he created during the eighteenth century.

The Christian tradition has had its own share of self-pro-claimed messiahs, often but not always associated with the imminence of the end of the world. In pre-medieval Europe, the 'Christ of Gevaudon' in France had an army of 3000 soldiers at his command, but was eventually hacked to death in 593 AD by men acting for the local bishop. A century and a half later, Aldebert of Soissons, who claimed to possess a letter from Jesus, declared himself a saint in his own lifetime and built up a following that was sufficiently large to worry Pope Zacchary far away in Rome. Twelfth-century Europe was a particularly good time for messiahs. Eudo de Stella declared himself to be the Son of God. Tanchelm of Antwerp, after a hugely successful preach-ing tour, began with the same modest claim (he even had an inner circle of twelve disciples) but later extended it to allege that he was God Himself – which is no doubt why he felt confi-dent enough to announce his betrothal to the Virgin Mary, a sacred statue standing in for his apparently otherwise engaged consort at a magnificent ceremony in front of a large congrega-tion of faithful followers.

Central Europe was a prolific producer of new sects in the late Renaissance and Post-Reformation period, including among many others the Taborites, Hussites, Anabaptists and Mennonites. Later, in the eighteenth and nineteenth centuries, Britain and the USA played host to myriad obscure and not-so-obscure cults, almost all centred round a single founder. Most, like the Shakers and the Oneidans, or the followers of 'Jumpin' Jesus' Matthews in New York or the Reverend Henry Prince in the more genteel surroundings of Victorian Brighton, faded away with the deaths of their founders, but others, like the Methodists, the Mormons and the Quakers, grew from strength to strength and acquired a momentum of their own.

Nor has the Islamic tradition been spared this phenomenon. In the aftermath of the Prophet Mohammed's death in 632 AD, factionalism developed around his various offspring and disciples, leading to the great split between the Sunni and Shi'ite branches (and the murder of the Imam Husain). The Shias looked forward to the appearance of a messiah, the Mahdi, who would lead them to final triumph over the Sunnis and the infidel. (The most famous recent claimant to this title was the Sudanese sheikh Mohammed Ahmed, who was responsible for causing General Gordon a little personal bother at Khartoum in 1885.) The ensuing millennium and a half has witnessed the emergence of numerous cults, large and small, each centred on a single messianic figure and each with its own traditions and beliefs. There were those that believed in reincarnation, those that believed in absolute sexual freedom, those that believed in the central importance of ecstasy brought on by pain or by mystical contemplation, even those that believed in ritual murder as a religious duty.

And this raises for us the central theme of this chapter: why is it that, uniquely within the Animal Kingdom, religion has such a stranglehold over our species? Why, despite our much vaunted intellect, do we repeatedly succumb to the demands of religious zealots – to the point where we are willing to lay down our own lives in the name of what everyone else believes to be an obvious fiction?

Worse still is the fact that we have so often been willing to slaughter tens of thousands of our fellow countrymen (never mind those of different race) merely because they happened to hold a different set of religious beliefs. Religion has been involved, either as the instigating force or, subsequently, as a justification, in enough of the conflicts that have dogged human history as to give us pause for thought. We cannot write such events off as insignificant examples of the occasional lunacy to which our species is sometimes prone any more than we were right to write off the behaviour of the Kasekela chimpanzee males as an odd aberration in an otherwise idyllic life in the forest.

This sense of community brings into focus the second pair of functions that religion seems to have, namely the coercive function that often spills over into state-based religion. In the context of small groups, religion seems to form a very strong basis for enforcing group norms, irrespective of what those norms happen to be. In small communities, such as those in which we have spent most of our existence as a species, religion probably served the crucial function of preventing individualistic interests undermining the effectiveness of coordinated group action. For species as intensely social as all primates are (and we humans have this trait in spades), the need to control the disruptive influence of those who buck the trend, the individualists (unless, of course, they happen to be charismatic leaders), the non-cooperators and the free riders, becomes paramount if the group is to do its job of ensuring that its individual members can survive and reproduce effectively. That pressure seems to have been responsible for producing a mind predisposed to acquiesce in the collective view, especially when this is expressed in strongly religious terms reinforced with an emotionally rich cocktail of music, dance and ritual. Religion, in short, probably evolved because it was such a good mechanism for bonding social groups and making them work together in the common interest. It was this, above all, that relentlessly drove the evolution of the religious mind.

A Solitary Candle Burning Bright

Theory of mind is crucial to the whole enterprise of religion. Religion, in its very basest form, requires us to suppose that there is a world other than the one we see, and that means second-order intentionality or theory of mind at the very least.

Since even the apes can only just aspire to that, it means that religion is unlikely to be found outside the immediate zoological family to which we belong. But I suspect that religion is actually more cognitively demanding than this.

In order to be able to engage in religious activities, I have to believe that there is a parallel world occupied by beings who have intentions that can be influenced by my prayers. In other words, I *believe* [1] that there are gods who *intend* [2] to influence my future. If these beings have intentions that I am unable to influence, then religion has no role to play: such beings are little different to the raging floods or erupting volcanoes that unexpectedly engulf us. A religion, if it is to have any real value, has to be able to influence the future for us.

But, second-order intentionality is not really enough to drive a metaphysical belief. If religion is to have any useful purpose, then these gods must be able to understand what *I* want. So it seems likely that religion must presuppose third-order intentionality: I *believe* [1] that there are gods who can be persuaded to *understand* [2] what I really *desire* [3] and who, having done so, will act on my behalf.

This, I think, is enough to explain the evolution of a religious sense – to provide the cognitive underpinnings for the personal sense of religion, my own particular beliefs and transcendental experiences. However, it is still not enough to explain the *communal* sense of religion, the large-scale phenomena of rituals and public commitment that are so central a part of religion as we practise it. Religion in its human form is nothing if it is not a *social* activity: we come together in common rituals and beliefs to form a community. To achieve that, I need at least fourth- (and maybe even fifth-) order intentionality: I *suppose* [1] that you *think* [2] that I *believe* [3] that there are gods who *intend* [4] to influence our futures (. . . because they understand our *desires* [5]?). Unless and until we come together in this way, we do not have religion, only personal belief. It is shared belief that makes religion what it is.

That being so, then it is perhaps obvious why humans – and only humans – seem to have religious systems. Only humans can aspire to fourth-order intentionality as a matter of course. More interestingly, only *some* humans can aspire to fifth- and sixth-order intentionality, which may explain why, among humans, only a relatively small number of individuals are successful religious leaders. Religious leaders, like good novelists, are a rare breed.

Dead Men's Tales

We are left, then, with one last question: when did religion first appear in human history? The short answer is that we have absolutely no idea. But some strands of evidence can perhaps give us some pointers.

We know that all living human societies have some form of religion, and this suggests that it is a common feature of the way the human mind is designed rather than something that, by some remarkable coincidence, has spontaneously evolved to have exactly the same form on a number of separate occasions in a number of different places. This suggests, in turn, that it has a modestly ancient origin that has little to do with the cultural diversity that has emerged in the last 30,000 years since the Upper Palaeolithic Revolution. The Upper Palaeolithic Revolution was confined pretty much to Europe and did not feature to anything like the same extent among our fellow modern humans in Africa, Asia and Australia. If so, then the trait must date back at least to the last common ancestor of all living humans. As we saw in Chapter 2, molecular genetics tells us that the last time the Eurasian and African branches of modern humans were united was around 70,000 years ago when the ancestors of the Eurasians left their African homeland. So the common roots of religion must lie in the interval between 70,000 years ago and the last common ancestor of all modern humans, sometime around 200,000 years ago. But how much earlier than that does a religious sense go? What archaeological evidence is there for religion?

Our problem is to know how to recognise the signature of religious belief in the archaeological record. After all, without knowing the oral history of Christianity, we would be unable to interpret the significance of crosses or chalices in Christian iconography. As with the problem of defining culture in animals, one solution is to look for phenomena that do not have any obvious functional use. The trouble is that most of these are also likely to have everyday uses and separating the everyday from the ritual may be tricky at this remove. Are Venus figures (those extraordinary Michelin-tyre female figurines that appear in the European archaeological record from about 30,000 years ago) fertility symbols (as some have assumed), images of goddesses, or just entertaining decorative art (the prehistoric equivalent of pin-ups)? There is, however, one facet of human behaviour that does provide us with evidence for belief in an afterlife, since it has one very concrete and relatively unmistakable form – burial.

The earliest uncontroversial evidence for deliberate burials comes from two Cro-Magnon sites at Predmostí and Dolní Vestonice in the Czech Republic, both dated to around 25,000 years ago. In one, two young men and a young woman were buried together, while in the other as many as eighteen individuals had been buried in a large pit covered by mammoth bones and limestone slabs. The Sungir site in Russia (dated to around 22,000 years ago) boasts the skeletons of two children placed head to head. One was covered in around 5000 beads whose positioning strongly suggests that they were part of the clothing the child had been buried in. In addition, some 250 perforated Arctic fox teeth encircled its waist as though they had once been attached to (or formed) a belt; an ivory animal pendant lay on its chest and an ivory pin near its throat. The other skeleton was associated with a similar number of beads arranged as though they had once been attached to clothing, and it too had an ivory pin at its throat. With the bodies was an assortment of large and small ivory lances, some antler 'wands', a sculpted ivory mammoth and the highly polished shaft of a human femur packed with red ochre (a substance often used as a decorative paint by prehistoric peoples, as well as by contemporary hunter-gatherer peoples in both southern Africa and Australia).

Although much has been made of putative graves at Neanderthal sites, some dating from as early as 50,000 years ago, the evidence in these cases is at best equivocal. There was a particular flurry of excitement a decade or so ago when flower pollen was identified from around a Neanderthal skeleton at Shanidar in modern Iraq. The presence of flower pollen, it was argued, implied flowers, and flowers could hardly have arrived there by accident: *ergo*, they must have been introduced as part of a funerary ritual. But the enthusiasm for Neanderthal religion waned somewhat when it was later pointed out that the burial site is heavily disturbed and the pollen could well have found its way into the 'grave' thanks to the activities of rodents, or been blown in by the wind long after the body had been deposited there.

Similarly enthusiastic claims were made for deliberate Neanderthal burials on the basis of the tools and animal bones that are often associated with their skeletons. One young boy at the Teshik-Tash site in southern Russia attracted particular interest because he was surrounded by half a dozen pairs of mountain goat horns. However, many archaeologists now believe that most of these associated bones and tools were probably not deliberately placed there: they are just part of the debris and detritus of Neanderthal living sites that accumulated, in some cases with the bodies of the dead, as time passed. Although the individuals in Neanderthal 'graves' are often in a fetal position (knees tucked up around the chin), a more prosaic explanation for this might simply be the desire to dig the smallest possible hole for the disposal of the body. Some Neanderthal bones even bear the distinctive marks of gnawing by hyenas and other carnivores, suggesting that there had been no attempt to bury the dead deliberately in such a way as to preserve them for an afterlife. Some even show signs (fine cut marks on the bone) suggesting that they may have been methodically defleshed – a feature that has been interpreted as evidence of cannibalism. In sum, it is fair to say that, even if the Neanderthals did bury their dead, their graves are just much less elaborate than those of the Cro-Magnons that replaced them.

On balance, the only conclusion we can draw from all this mortuary evidence is that a real sense of an afterlife to which the dead might journey – and where they might need the accoutrements of everyday life to ease their way – appeared only with the Cro-Magnon-age peoples of the later Palaeolithic. And that would have been a *very* long time after language first evolved. This is pretty much supported by the evidence from prehistoric art. Female figurines (Venus figures) and carvings of animals (in some cases, engravings on bone or ivory) have been found in about thirty caves scattered across southern Europe from Spain to southern Russia, mostly associated with dates in the region of 28–21,000 years ago. In addition, some 150 caves containing prehistoric art have been found (almost all in southern France and northern Spain, though a few are known from southern Germany and further to the east). The oldest (the Chauvet cave in the Ardèche valley in France) is dated to around 31,000 years ago. These cave paintings are all associated with the Magdalenian peoples (the later Cro-Magnons) who succeeded the Neanderthals in Europe.

The purpose of all this artwork remains obscure, but the fact that it is often found deep underground in places that would have been extremely difficult to gain access to has been interpreted as indicating a quasi-religious function or some kind of ritual function (associated perhaps with puberty or hunting rituals). Certainly, the subjects that most attracted the artists' minds seem to be those associated with animals. The fact that many of the rituals of living hunter-gatherers such as the !Kung San of southern Africa are also associated with animal magic perhaps lends some support to this suggestion. The principal puberty ritual of the San people, for example, is the so-called 'eland dance' in which the dancers wear eland-skin cloaks and head-dresses and give their blood for the ritual.

But maybe this can only tell us when a belief in a particular kind of afterlife (one where one's own body and possessions were needed) first evolved. Perhaps people prior to that had been just as religious, but did not associate their earthly bodies with where their spirits continued to live after death. After all, it was manifestly the case that they did not take their physical bodies with them, wherever they went at death. Moreover, by no means all contemporary or historical religions regard the preservation of the corpse as essential: some (like the Hindus and many of the Indo-European groups) cremate it, others (like the Parsees) expose it to carrion-eaters. If burial is not necessarily a certain marker for a belief in an afterlife, the archaeological record may not be especially helpful.

An alternative possibility might be to consider the cognitive requirements, just as we did in exploring the time at which language might have evolved. If religion requires fourth-, or even fifth-, order intentionality, then we should be able to exploit the relationship we found in Chapter 3 between levels of intentionality, brain size and the fossil record to see when the requisite fifth-order necessary to support religion as a communal activity might have appeared. Figure 6 shows the pattern that emerges when we plot the resulting highest achievable levels of intentionality against the age of the fossil hominid populations.

Taken at face value, the results shown in Figure 6 suggest that although third-order intentionality would have characterised *Homo erectus*, fourth-order intentionality would not have made its appearance until sometime around 500,000 years ago when archaic humans came on the scene. If religion requires fourth-

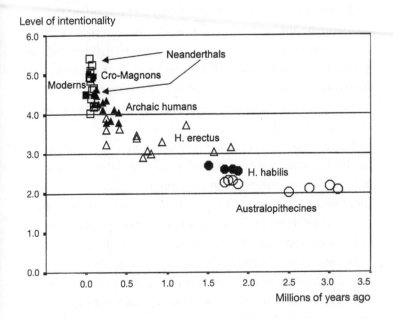

Figure 6: Given that we can estimate the brain volumes of fossil hominids (see Figure 2, page 29), we can use standard equations to estimate the size of their frontal lobes, and then use the relationship between frontal-lobe volume and achievable intentionality level found in monkeys, apes and modern humans to estimate the likely level of mind-reading that each fossil hominid population might have achieved. Each point represents the mean value for one population of fossil hominids.

order intentionality, then it might have coincided with the rise of language which seems to have appeared at this point (see Figure 5, page 125). This is probably not too surprising, because the communal nature of religion depends on language: language is needed to explain the religious system so as to persuade others to adopt it, so it would have had to be in place before it could be used to create religion. Fifth-order intentionality, however, did not appear until much later, being associated with anatomically modern humans (the Cro-Magnons and modern humans). If fifth-order is needed to sustain religion as a *communal* enterprise, then in all likelihood it probably dates back only 200,000 years at most. It may be no accident that this seems to be the same time frame for the appearance of full *grammatical* language (at least in so far as the evidence for 'genes' for grammar can tell us anything). Full grammatically structured language is essential for the transmission of the metaphysical concepts that necessarily underpin transcendental religious beliefs.

There is, however, an interesting question hovering quietly in the background: the later Neanderthal populations had brains as large (or even larger) than ours and if the logic of this argument holds, then we might expect religion to have been a feature of their lives too. There are three possible positions we can take, though we cannot really decide between them on the evidence we have at present. One is that, if large brain size (and hence fifth-order intentionality) evolved independently among the Neanderthals and the Cro-Magnons, then both sub-species may have evolved a religious approach to life independently of each other. Alternatively, since religion essentially represents a *software* and not a hardware change, it is possible that despite the possession of fifth-order intentionality, the Neanderthals failed to develop religion, at least in some of its key social aspects. Religion would thus have been a serendipitous cultural invention of one particular (and unusually thoughtful?) anatomically modern human somewhere on an African plain. That would fit well with our experience of how new religions or sects arise. But religion has the interesting capacity to spread like wildfire through a population, so once it had appeared as a kind of cultural mutation somewhere, then its spread through neighbouring (and eventually far distant) communities would have been very rapid. The third possibility is that the Neanderthal brain was organised in a different way from that of modern humans, so that although they had a larger total brain volume than modern humans, much of this extra volume was in the visual areas at the back of the brain (hence the famous Neanderthal 'bun') with proportionately less in the frontal lobe. If they did indeed have a smaller frontal lobe, then their achievable levels of intentionality would have been lower – perhaps low enough to preclude the development of full-blown communal religion.

The last possibility might provide an explanation for that most vexing of all archaeological conundrums: why, despite the fact that they had previously been so successful, did the Neanderthals die out so rapidly after the arrival of the Cro-Magnons in Europe? One answer might be the added power that religion as a coercive force gave to Cro-Magnon groups, allowing them to act more cohesively and effectively at the socio-political level when forced into ecological competition with Neanderthals. Without religion as a device for bonding social groups, they may have been no match for the invaders from Africa who were our direct ancestors.

The Shaman's Vision

We are apt to think of religion in terms of the great international organised religions of the modern world (Hinduism,

Jainism, Buddhism, Sikhism, Shinto, Islam, Judaism and Christianity) or the great religions of the historical past (the sun-worship of the Aztecs and Incas, the pantheism of Greek and Roman classical state religion, the Zoroastrianism of ancient Persia – probably the oldest organised religion in the world, placed on a firm footing by its first prophet Zarathustra around 1200 BC, and now represented mainly by the Parsees of western India). One way or another, all of these are characterised by philosophically sophisticated systems of thought, international structures of bureaucracy and highly organised forms of worship, often in specially constructed (and, in many cases, lavishly ornamented) buildings. But it has not always been thus. Perhaps the fact that, within these great religions of the world, new movements are constantly being born in people's houses, in open fields, or in village halls, should remind us that religion has an intimacy that stems, not from the political hierarchies of priests, bishops, presbyters and popes, but from the intimacies of personal relationships between small groups of people.

And it was thus, perhaps, that religion first started among the roving bands of our hunter-gatherer ancestors. The nearest that we can come to this are the traditional religions of living hunter-gatherers and other small-scale tribal peoples. Among the !Kung San of the Kalahari desert in southern Africa, religion is expressed in belief systems about the hidden spirit world and in the rituals of the trance dances that give humans access to that world. There are no priests, even though some individuals may be regarded as particularly adept at the business of communicating with the spirit world. We can perhaps use the term *shaman* to refer to these individuals, even though strictly speaking this term is associated with the particular set of beliefs and rituals of the Siberian peoples whose term this is. In at least some cases, these tribal religions do not even seem to have the concept of an afterlife to which we go when we die. This seems to be true not only of the !Kung San hunter-gatherers of the Kalahari desert in southern Africa, but also of the Masai pastoralists of East Africa.

David Lewis-Williams, a South African archaeologist, has argued that there is good evidence for believing that shamanism was the original form of religion among prehistoric humans. One line of evidence is the ubiquity in all human societies of the ability to enter trance states – sometimes induced by music and dance, sometimes by special meditative practices, but occasionally even by the use of psychotropic drugs like the mescaline favoured by the Mexicans. Another reason why Lewis-Williams thinks that shamanism may be the primordial form of religion is the fact that many of the abstract elements in prehistoric cave paintings, as well as the rock art of contemporary hunter-gatherers in southern Africa and Australia, involved patterns of dots, grids, zigzags and meandering lines that bear an uncanny resemblance to the experiences described by people who take hallucinogens during scientific experiments. They experience pinpoints of light or lines that flicker and pulsate with an intensity and brilliance that leaves the mind overwhelmed by the experience. In the final stages of these experiences, and especially so in cultures that are predisposed to see the world this way, individuals may feel themselves floating outside their own bodies, sometimes even turning into a specific animal or mythical form. In deistic cultures like the Christian tradition, those who enter trances may feel that they are gradually being absorbed into the godhead itself.

Lewis-Williams argues that it is these experiences that the rock artists are trying to capture. When the San rock art of southern Africa shows humans, it often shows them in lines, frequently carrying sticks. These have been mistaken for lines of men out hunting or, perhaps, going into battle carrying spears. Lewis-Williams suggests that they may in fact be depictions of trance dances. One reason for thinking this is that there are sometimes groups of women (evident by their breasts or the aprons or leather skirts they are wearing) in the background, or even intermingled in the line of men itself. A second reason is the presence of therianthropes (human figures with animal heads). These hardly seem relevant to hunting magic, let alone battles, but they do seem to be a common feature of trance states. In addition, sometimes the male figures have what are either sticks or blood coming out of their noses, much as the noses of San trance dancers drip blood at the height of their dance when they finally enter into a trance state.

This expression of a world-beyond-the-world that, in reality, lies within one's very head carries enormous potency. It is not difficult to see how it might have originated from the accidental experiences of a few individuals engaged in communal bonding activities made possible by music and dance. Being able to master these experiences, to bring them on at will, and then to lead others through them provides adepts with enormous charisma and power. The experiences they involve have the frisson of fear that is inevitably associated with the unknown that is beyond control. Being led through them by an adept creates a sense of excitement tinged with that essential degree of confidence that one will survive the danger. This is a powerful and heady mix, more than enough to turn a man's mind.

But the very seeds of that situation already contain within it the basis for developing institutional religions. The shaman becomes a holy man or woman, someone with magical powers who is able to control this world as well as the world beyond, someone who can work miracles on behalf of the earth-bound, provide comfort to the living and speed the dead on their way to whatever lies beyond. We are only a step away from priests and hierarchies, and the paraphernalia of institutions.

In terms of the origins of religion, however, the story that I have sketched out here suggests that the earliest stages may well have been very personal and intimate. Perhaps they were brought on by music and dance (in which case, they might well predate the appearance of *Homo sapiens*, though it seems very unlikely that they would go all the way back to the origins of *Homo erectus* two million years ago). Their efficacy in social bonding (generated by the surges in endorphins brought on by trances) is likely to have been their starting point – a purely chemical effect that helped to bond large dispersed hunter-gatherer groups. Only much later would the intellectual advantages of religion have become evident. Their role in stabilising the universe, providing a unifying set of cultural beliefs to provide a flagstaff around which group members could nail their individual colours – and, eventually, a means of enforcing adherence to the group's norms of behaviour – would all have emerged much later.

This must have been so because, when you analyse its cognitive demands as we have done, fully social religion minimally requires fourth-order intentionality (to understand it) and perhaps fifth-order to create it. Religion as we know it in its communal form could not have evolved before humans acquired fifth-order intentionality (which appears only with anatomically modern humans around 200,000 years ago) and language (which evolved some time between 500,000 and 200,000 years ago).

As remarkable as our achievements in the arts and sciences may be, it is hard to escape the conclusion that religion is the one phenomenon in which we humans really are different in some qualitative sense from our ape cousins. In most other respects, we can argue a convincing case for humans just being apes on a grander scale. But religion represents a genuine shift of gear into a new dimension that raises us into another world above and beyond the experiences of our ape cousins. That will, no doubt, bring comfort to some. But it is, in the end, something of a double-edged sword. Religion has also been the source of some of our worst nightmares.

The eighteenth-century French philosopher and mathematician René Descartes left us with a legacy that has not been easy to shake off. In a deliberate attempt to prove the existence of God, he reinforced the gulf between animals and ourselves. Not only did he thereby provide a legitimacy for how we mistreat the rest of the planet, but he also left us with an excessively inflated view of ourselves. Of course, Descartes was surely right to emphasise just how different humans are from other animals. We genuinely *are* different, not least in several key psychological respects. It is these differences that have allowed us to evolve that handful of features – language, culture, religion and science – that really do mark us off from the other animals with whom we have been privileged to share so much of our history. Those features allow us to have an extraordinarily rich mental life that is, so far as we can tell, genuinely unique.

Yet we should, at the same time, see these seemingly remarkable phenomena in proper perspective. Examined close up, they are simply the emergent properties of some very basic biological and psychological processes that we share with most of our cousins among the primates. The difference is simply the scale on which we can exercise these capacities.

Accidents of history placed exacting demands on our predecessors. Many of their contemporaries failed to meet those challenges and left no descendants, but the few that did sent our history spinning down unexpected channels at key moments. Their responses to the exigencies of the moment in those desperate battles to survive and reproduce successfully were as much a part of their primate biology as anything their ancestors had ever done before them. We can, perhaps, pinpoint in time the moment when any one component of our nature first appeared. Yet, there was no one point at which we can say that 'this was when we were set apart', no grand moment of conversion on the road to Damascus that made the non-human suddenly human. Rather, what we see is the gradual accumulation of those key components one by one, each a response to some unique circumstance, some particular challenge, but each paving the way for the next in the long sequence that, ultimately, led us to where we now are.

History has handed us a rare and privileged plate. Honesty would have us accept that we have sometimes used those remarkable capacities for anything but benign purposes. Religion, no less than anything else, has had a particularly black history. Yet it might be wrong to conclude from this that we must do away with religion altogether. We should not, in our haste, overlook the important role religion has played in human affairs, helping to bond communities and so enabling them to meet the challenges that the planet has thrown at them. Even today, its contribution to human psychological wellbeing is probably sufficient to raise serious questions about whether the human race could do without it.

In a rational humanistic world, such as that which Descartes himself set in train, our natural response must be to wean ourselves off the drug that religion ultimately is. But to succeed in this, we will need to find something in the social sphere to replace it. As Robert Putnam has pointed out in his book *Bowling Alone*, there is a great deal of evidence to suggest that well integrated communities (and this means ones that are internally well connected and have a sense of communal belonging, often generated by active social and religious institutions) suffer from less antisocial behaviour and crime – no doubt in part because of internal policing, but also partly because of the sense of obligation and community that adherence to common values and beliefs engenders. The problem for the contemporary rationalist is how to recreate this sense of community without resorting to the mechanism of religion, because religion works its will most effectively when we abandon rational thought and surrender ourselves to the mysterious and the ineffable.

We are, you might say, an oddly mixed-up species, an evolutionary Heath-Robinson of a job. But then, as evolutionary biologists are never slow to point out, that is what evolution is all about: evolution does not set out to produce engineering perfection, but rather simply adapts what is already there to do a novel job as best it can when the need arises. Nor does evolution come for free: any change in design that bears a benefit inevitably incurs a cost. The processes of evolution simply lead to where the benefits of any given change outweigh the costs. So it is that we are a hotch-potch of things that seemed like a good idea at the time, but which, with hindsight, might perhaps have been done better or differently. In that respect, we are no different from any of the other species that has ever lived. Our challenge, as it has always been, is to live with our imperfections, yet leave the world a better place than we found it.

Bibliography

Chapter 1 Visions in Stone

Lewis-Williams, D. (2002). *The Mind in the Cave*. Thames and Hudson, London.
Lewis-Williams, D. (2002). *A Cosmos in Stone: Interpreting Religion and Society Through Rock Art*. Altamira Press, New York.

Chapter 2 The Ape on Two Legs

Aiello, L. C. (1993). The fossil evidence for modern human origins in Africa: a revised view. *American Anthropologist* 95: 73–96.
Aiello, L. C. and Wheeler, P. (1995). The expensive tissue hypothesis: the brain and the digestive system in human evolution. *Current Anthropology* 36: 199–221.
Diamond, J. (1991). *The Rise and Fall of the Third Chimpanzee*. Random House, London.
Diamond, J. (1998). *Guns, Germs and Steel: A Short History of Everybody for the last 13,000 Years*. Random House, London.
Fleagle, J. G. (1999). *Primate Adaptation and Evolution*. 2nd edition. Academic Press, New York.
Ingman, M., Kaessmann, H., Pääbo, S. and Gyllensten, U. (2000). Mitochondrial genome variation and the origin of modern humans. *Nature* 408: 708–713.
Klein, R. (1999). *The Human Career*. 2nd edition. University of Chicago Press, Chicago.
Krings, M., Stone, A., Schmitz, R. W., Krainitzki, H., Stoneking, M. and Pääbo, S. (1997). Neandertal DNA sequences and the origin of modern humans. *Cell* 90: 19–30.

Lahr, M. M. and Foley, R. (1994). Multiple dispersals and modern human origins. *Evolutionary Anthropology* 3: 48–60.

Mellars, P. and Stringer, C. (eds) (1989). *The Human Revolution: Behavioural and Biological Perspectives on the Origins of Modern Humans.* Edinburgh University Press, Edinburgh.

Stoneking, M. (1993). DNA and recent human evolution. *Evolutionary Anthropology* 2: 60–73.

Stringer, C. and Gamble, C. (1993). *In Search of the Neanderthals: Solving the Puzzle of Human Origins.* Thames and Hudson, London.

Tattersall, I. (1999). *The Last Neanderthal: The Rise, Success and Mysterious Extinction of Our Closest Human Relatives.* Westview Press, New York.

Chapter 3 Mental Magic

Astington, J. W. (1993). *The Child's Discovery of the Mind.* Cambridge University Press, Cambridge.

Barrett, L., Dunbar, R. I. M. and Lycett, J. E. (2002). *Human Evolutionary Psychology.* Palgrave, Basingstoke and Princeton University Press, Princeton, NJ. (See especially chapter 10.)

Baron-Cohen, S. (2003). *The Essential Difference: Men, Women and the Extreme Male Brain.* Allen Lane, Harmondsworth.

Baron-Cohen, S. and Hammer, J. (1997). Is autism an extreme form of the 'male brain'? *Advances in Infancy Research* 11: 193–217.

Boroditsky, L. (2000). Metaphoric structuring: understanding time through spatial metaphors. *Cognition* 75: 1–28.

Byrne, R. (1995). *The Thinking Ape: The Evolutionary Origins of Intelligence.* Oxford University Press, Oxford.

Byrne, R. W. and Whiten, A. (eds) (1988). *Machiavellian Intelligence: Social Expertise and the Evolution of Intellect in Monkeys, Apes and Humans.* Oxford University Press, Oxford.

Boysen, S. T. and Berntson, G. G. (1995). Responses to quantity: perceptual versus cognitive mechanisms in chimpanzees (*Pan troglodytes*). *Journal of Experimental Psychology: Animal Behavior Processes* 21: 82–86.

Cheney, D. and Seyfarth, R. M. (1980). *How Monkeys See the World.* Chicago University Press: Chicago.

Dunbar, R. I. M. (1996). *Grooming, Gossip and the Evolution of Language.* Faber and Faber, London and Harvard University Press, Cambridge, Mass.

Dunbar, R. I. M. (1998). The social brain hypothesis. *Evolutionary Anthropology* 6: 178–190.

Dunbar, R. I. M. (2000). Causal reasoning, mental rehearsal and the evolution of primate cognition. In: C. Heyes and L. Huber (eds) *The Evolution of Cognition*, pp. 205–231. MIT Press, Cambridge, Mass.

Dunbar, R. I. M. (2002). Why are apes so smart? In: P.Kappeler and M.

Peirera (eds) *Primate Life Histories*. MIT Press, Cambridge, Mass.

Happé, F. (1994). *Autism: An Introduction to Psychological Theory*. University College London Press, London.

Hare, B., Call, J., Agnetta, B. and Tomasello, M. (2000). Chimpanzees know what conspecifics do and do not see. *Animal Behaviour* 59: 771–785.

Joffe, T. H. (1997). Social pressures have selected for an extended juvenile period in primates. *Journal of Human Evolution* 32: 593–605.

Joffe, T. and Dunbar, R. I. M. (1997). Visual and socio-cognitive information processing in primate brain evolution. *Proceedings of the Royal Society London, B*, 264: 1303–1307.

Kinderman, P., Dunbar, R. I. M. and Bentall, R. P. (1998). Theory-of-mind deficits and causal attributions. *British Journal of Psychology* 89: 191–204.

Kudo, H. and Dunbar, R. I. M. (2001). Neocortex size and social network size in primates. *Animal Behaviour* 62: 711–722.

Lewis, K. (2001). A comparative study of primate play behaviour: implications for the study of cognition. *Folia Primatologica* 71: 417–421.

Mitchell, P. (1997). *Introduction to Theory of Mind*. Arnold, London.

Mithen, S. (1996). *The Prehistory of the Mind*. Thames and Hudson, London.

Pawlowski, B., Lowen, C. L. and Dunbar, R. I. M. (1998). Neocortex size, social skills and mating success in primates. *Behaviour* 135: 357–368.

Povinelli, D. (1999). *Folk Physics for Apes*. Oxford University Press, Oxford.

Tomasello, M. (2001). *The Cultural Origins of Human Cognition*. Harvard University Press, Cambridge, Mass.

Tomasello, M. and Call, J. (1997). *Primate Social Cognition*. Oxford University Press, Oxford.

de Waal, F. (1982). *Chimpanzee Politics: Power and Sex Among the Apes*. Unwin, London.

Whiten, A. and Byrne, R.W. (1988). Tactical deception in primates. *Behavioral and Brain Sciences* 11: 233–273.

Wozniak, R. (2003). Oskar Pfungst: *Clever Hans (The Horse of Mr von Osten)*. http://www.thoemmes.com/psych/pfungst.htm

Phineas Gage: http://www.hbs.deakin.edu.au/gagepage

Chapter 4 Brother Ape

Barrett, L., Dunbar, R. I. M. and Lycett, J. E. (2002). *Human Evolutionary Psychology*. Palgrave, Basingstoke and Princeton University Press, Princeton, NJ. (See especially chapter 8.)

Campbell, Ann. (2000). *A Mind of Her Own: The Evolutionary Psychology of Women*. Oxford University Press, Oxford.

Crook, J. H. (1997). The indigenous psychiatry of Ladakh. Part I. Practice theory approaches to trance possession in the Himalayas. *Anthropological Medicine* 4: 289–307.

Crook, J. H. and Crook, S. J. (1988). Tibetan polyandry: problems of adaptation and fitness. In: L. Betzig, M. Borgerhoff-Mulder and P. Turke (eds), *Human Reproductive Behaviour*, pp 97–114. Cambridge University Press, Cambridge.

Dickeman, M. (1979). Female infanticide, reproductive strategies and social stratification: a preliminary model. In: N. A. Chagnon and W. Irons (eds), *Evolutionary Biology and Human Social Behaviour*, pp. 321–367. Duxbury Press, North Scituate, Mass.

Dunbar, R. I. M. (1988). *Primate Social Systems*. Chapman and Hall, London.

Edgerton, R. B. (1992). *Sick Societies: Challenging the Myth of Primitive Harmony*. Free Press, New York.

Goodall, J. (1971). *In the Shadow of Man*. Weidenfeld and Nicolson, London.

Harcourt, A. H. and Greenberg, J. (2001). Do gorilla females join males to avoid infanticide: a quantitative model. *Animal Behaviour* 62: 905–915.

Hrdy, S. B. (1999). *Mother Nature*. Harvard University Press, Cambridge, Mass.

Jackson, M. (ed) (2002). *Infanticide: Historical Perspectives on Child Murder and Concealment, 1550–2000*. Ashgate, Aldershot.

Panter-Brick, C. and Smith, M. T. (eds) (2000). *Abandoned Children*. Cambridge University Press, Cambridge.

van Schaik, C. P. and Dunbar, R. I. M. (1990). The evolution of monogamy in large primates: a new hypothesis and some critical tests. *Behaviour* 115: 30–62.

van Schaik, C. P. and Janson, C. H. (eds) (2000). *Infanticide by Males and its Implications*. Cambridge University Press, Cambridge.

Strassman, B. I. and Dunbar, R. I. M. (1999). Human evolution and disease: putting the Stone Age in perspective. In: S. C. Stearns (ed) *Evolution in Health and Disease*, pp. 91–101. Oxford University Press, Oxford.

Symonds, D. A. (1997). *Weep Not for Me: Women: Ballads and Infanticide in Early Modern Scotland*. Pennsylvania State University Press, University Park.

Voland, E. (1989). Differential parental investment: some ideas on the contact area of European social history and evolutionary biology. In: V. Standen and R. A. Foley (eds) *Comparative Socioecology: the Behavioural Ecology of Humans and Other Mammals*, pp. 391–403. Blackwell, Oxford.

de Waal, F. (1982). *Chimpanzee Politics: Power and Sex Among the Apes*. Unwin, London.

Wrangham, R. W. and Peterson, D. (1997). *Demonic Males: Apes and the Origins of Human Violence*. Bloomsbury, London.

Chapter 5 So Sweetly Sung in Tune

Aiello, L. C. and Dunbar, R. I. M. (1993). Neocortex size, group size and the evolution of language. *Current Anthropology* 34: 184–193.

Barrett, L., Dunbar, R. I. M. and Lycett, J. E. (2002). *Human Evolutionary Psychology*. Palgrave, Basingstoke and Princeton University Press, Princeton, NJ. (See especially chapter 12.)

Deacon, T. (1997). *The Symbolic Species: The Coevolution of Language and the Human Brain*. Allen Lane, Harmondsworth.

Dunbar, R. I. M. (1996). *Grooming, Gossip and the Evolution of Language*. Faber and Faber, London and Harvard University Press, Cambridge, Mass.

Goldstein, A. (1980). Thrills in response to music and other stimuli. *Physiological Psychology* 8: 126–129.

Juslin, P. N. and Sloboda, J. A. (eds) (2001). *Music and Emotion: Theory and Research*. Oxford University Press, Oxford.

Kay, R. F., Cartmill, M. and Barlow, M. (1998). The hypoglossal canal and the origin of human vocal behavior. *Proceedings of the National Academy of Sciences, USA*, 95: 5417–5419.

MacLarnon, A. M. and Hewitt, G. P. (1999). The evolution of human speech: the role of enhanced breathing control. *American Journal of Physical Anthropology* 109: 341–363.

Nettle, D. (1999). *Linguistic Diversity*. Oxford University Press, Oxford.

Provine, R. (1997). *Laughter: A Scientific Investigation*. Faber and Faber, London.

Seepersand, F. (1999). *Laughter and Language Evolution: Does the Topic of Conversation Eliciting the Most Laughter Last Longer?* MSc thesis, University of Liverpool.

Shami, P. and Stuss, D. T. (1999). Humour appreciation: a role for the right frontal lobe. *Brain* 122: 657–666.

Smith, E. A. (1991). *Inujjuamiut Foraging Strategies*. Aldine, New York.

Stowe, J. (2000). *Investigation into the Possible Influence of Laughter on Endorphin Release through Pain Tolerance*. MSc thesis, University of Liverpool.

Wallin, N. L., Merker, B. and Brown, S. (eds) (2000). *The Origins of Music*. MIT Press, Cambridge, Mass.

Chapter 6 High Culture

Barrett, L., Dunbar, R. I. M. and Lycett, J. E. (2002). *Human Evolutionary Psychology*. Palgrave, Basingstoke and Princeton University Press, Princeton, NJ. (See especially chapter 13.)

Boesch, C. and Tomasello, M. (1998). Chimpanzee and human cultures. *Current Anthropology* 39: 591–614.

Dunbar, R. I. M., Knight, C. D. and Power, C. (eds) (1999). *The Evolution of Culture*. Edinburgh University Press, Edinburgh.

McGrew, W. (1992). *Chimpanzee Material Culture: Implications for Human Evolution*. Cambridge University Press, Cambridge.

Morris, D. (1977). *Manwatching: A Field Guide to Human Behaviour*. Grafton, London.

Sperber, D. (1996). *Explaining Culture: A Naturalistic Approach*. Blackwell, Oxford.

Tomasello, M. (2001). *The Cultural Origins of Human Cognition*. Harvard University Press, Cambridge, Mass.

Tomasello, M., Kruger, A. and Ratner, H. (1993). Cultural learning. *Behavioral and Brain Sciences* 16: 450–488.

Whiten, A., Goodall, J., McGrew, W. C., Nishida, T., Reynolds, V., Sugiyama, Y., Tutin, C. E. G., Wrangham, R. W. and Boesch, C. (1999). Culture in chimpanzees. *Nature* 399: 682–685.

Chapter 7 Thus Spake Zarathustra

Anon. (1994). *The World's Religions*. Lion Books, Oxford.

d'Aquili, E. and Newberg, A. (1999). *The Mystical Mind: Probing the Biology of Religion*. Fortress Press, Minneapolis.

Armstrong, K. (2000). *The Battle for God*. HarperCollins, London.

Beit-Hallahmi, B. and Argyle, M. (1997). *The Psychology of Religious Behaviour, Belief and Experience*. Routledge, London.

Boyer, P. (2001). *Religion Explained: The Human Instincts that Fashion Gods, Spirits and Ancestors*. Weidenfeld and Nicolson, London.

Cohn, N. (1970). *The Pursuit of the Millennium: Revolutionary Millenarians and Mystical Anarchists of the Middle Ages*. Oxford University Press, Oxford.

Crook, J. H. and Low, L. (1997). *The Yogins of Ladhak*. Motilal Banarsidass, Delhi.

Dunbar, R. I. M. (1995). *The Trouble with Science*. Faber and Faber, London and Harvard University Press, Cambridge, Mass.

Flinn, M. V. and England, B. (1995). Childhood stress and family environment. *Current Anthropology* 36: 854–866.

Frankel, B. G. and Hewitt, W. E. (1994). Religion and well-being among Canadian university students: the role of faith groups on campus. *Journal of the Scientific Study of Religion* 33: 62–73.

Grayson, D. K. (1993). Differential mortality and the Donner Party disaster. *Evolutionary Anthropology* 2: 151–159.

Hamilton, M. (2001). *The Sociology of Religion*. 2nd edn. Routledge, London

Hinde, R. A. (2000). *Why Gods Persist*. Routledge, London.

House, J. S., Umberson, D. and Landis, K. R. (1988). Structure and processes of social support. *Annual Review of Sociology* 14: 293–318.

Kaplan, R. H. and Toshima, M. T. (1990). The functional effects of social relationships on chronic illness and disability. In: B. R. Sarason (ed), *Social Support: An Interactional View*. Wiley, New York

Klein, R. (1999). *The Human Career*. 2nd edition. University of Chicago Press, Chicago.

Knight, C. D. (1999). *Blood Relations: Menstruation and the Origins of Culture*. Yale University Press, New Haven, Conn.

Koenig, H. G. and Cohen, H. J. (eds) (2002). *The Link Between Religion and Health: Psychoneuroimmunology and the Faith Factor*. Oxford University Press, Oxford.

Levin, J. S. (1994). Religion and health: is there an association, is it valid, and is it causal? *Social Science and Medicine* 38: 1475–1482.

Lewis-Williams, D. (2002). *The Mind in the Cave*. Thames and Hudson, London.

Lewis-Williams, D. (2002). *A Cosmos in Stone: Interpreting Religion and Society through Rock Art*. Altamira Press, New York.

McCullogh, J. M. and York Barton, E. (1991). Relatedness and mortality risk during a crisis year: Plymouth colony, 1620–1621. *Ethology and Sociobiology* 12: 195–209.

Muncy, R. L. (1973). *Sex and Marriage in Utopian Communities: 19th Century America*. Indiana University Press, Bloomington.

Newberg, A., d'Aquili, E. and Rause, V. (2001). *Why God Won't Go Away*. Ballantine Books, New York.

Putnam, R. D. (2000). *Bowling Alone: The Collapse and Revival of American Community*. Simon and Schuster, New York.

Rouget, G. (1985). *Music and Trance: A Theory of the Relations Between Music and Possession*. University of Chicago Press, Chicago.

Sherratt, A. (1991). Sacred and profane substances: the ritual of narcotics in later Neolithic Europe. In: P. Garwood et al (eds), *Sacred and Profane*, pp. 50–64. Oxford Committee for Archaeology, Oxford.

Spence, J. (1954). *One Thousand Families in Newcastle*. Oxford University Press, Oxford.

Strassman, B. I. and Dunbar, R. I. M. (1999). Human evolution and disease: putting the stone age into perspective. In: S. C. Stearns (ed), *Evolution in Health and Disease*, pp. 91–101. Oxford University Press, Oxford.

Wilson, C., Wilson, D. and Wilson, R. (1992). *Cults and Fanatics*. Magpie Books, London.

Index